T0338632

SLIDING-ROLLING CONTACT AND IN-HAND MANIPULATION

SLIDING-ROLLING CONTACT AND IN-HAND MANIPULATION

Lei Cui
Curtin University, Australia

Jian S. Dai
King's College London, UK

NEW JERSEY • LONDON • SINGAPORE • BEIJING • SHANGHAI • HONG KONG • TAIPEI • CHENNAI • TOKYO

Published by

World Scientific Publishing Europe Ltd.
57 Shelton Street, Covent Garden, London WC2H 9HE
Head office: 5 Toh Tuck Link, Singapore 596224
USA office: 27 Warren Street, Suite 401-402, Hackensack, NJ 07601

Library of Congress Cataloging-in-Publication Data
Names: Cui, Lei (Mechanical engineer), author. | Dai, Jian S, author.
Title: Sliding–rolling contact and in-hand manipulation / Lei Cui,
Curtin University, Australia, Jian S Dai, King's College London, UK.
Description: New Jersey : World Scientific, 2020. | Includes bibliographical references and index.
Identifiers: LCCN 2020007882 | ISBN 9781786348425 (hardcover) | ISBN 9781786348432 (ebook) |
ISBN 9781786348449 (ebook other)
Subjects: LCSH: Robots--Control systems. | Robot hands--Control systems. |
Sliding mode control. | Rolling contact. | Geometry, Differential.
Classification: LCC TJ211.35 .C85 2020 | DDC 629.8/95--dc23
LC record available at https://lccn.loc.gov/2020007882

British Library Cataloguing-in-Publication Data
A catalogue record for this book is available from the British Library.

For any available supplementary material, please visit
https://www.worldscientific.com/worldscibooks/10.1142/Q0249#t=suppl

Desk Editors: Ramya Gangadharan/Michael Beale/Shi Ying Koe

Typeset by Stallion Press
Email: enquiries@stallionpress.com

Printed in Singapore

Lei Cui would like to dedicate this book with love and affection to his wife Qin Yu and his son Haoqian Cui

Preface

Robots interact with the world through curves and surfaces. Differential geometry is a branch of mathematics that uses ideas and techniques from differential calculus to study the geometry of curves and surfaces. Differential geometry is a beautiful subject, providing some of the most visually appealing geometric objects, for example, Mobius strip and Klein bottle, as well as intuition and insight into other subjects, for example, those in the excellently crafted book titled *Geometry and the Imagination* by David Hilbert and S. Cohn-Vossen (AMS Chelsea Publishing).

This book applies the moving-frame method, which was developed extensively by Élie Cartan, one of the greatest differential geometers, and the adjoint approach, which was conceived by Ernesto Cesàro, an insightful Italian differential geometer, to studying the kinematics of two surfaces subject to rolling contact and sliding–rolling contact. The book differs from other books on differential geometry in the following respects. First, the book deals with two surfaces, and the geometry of both surfaces comes into play. Second, the book focuses on the geometry of the two surfaces within the encompassing space (extrinsic) rather than within the surfaces (intrinsic) because the book is concerned with the kinematics of one surface in three-dimensional Euclidean space, the world.

The book consists of the following chapters. Chapter 1 covers the aims and objectives of the book and also describes the organization of the contents in the book. Chapter 2 presents the ideas of the moving-frame method and the application of this method to curves and surfaces. Chapter 3 introduces the adjoint approach, which is a powerful tool to deal with two geometric entities and the fixed-point conditions. Chapter 4 embeds rolling-contact constraints into the Darboux frames respectively on the two contact trajectories and uses the adjoint approach to obtain the kinematics of the moving surface. Chapter 5 considers sliding–rolling contact and uses the adjoint approach to obtain the induced angular velocity by the sliding motion. Chapter 6 solves the problem of inverse kinematics of rolling contact when the velocity of the moving surface is given. Chapter 7 extends the work in Chapter 5 and solves the problem of inverse kinematics of sliding–rolling contact. Chapter 8 presents a framework of in-hand manipulation of a robotic hand with an articulated palm — the Metahand — when the fingertips and the object maintain fixed-point contact. Chapter 9 applies the concept of the Gauss map to investigating the palm workspace and posture workspace of the Metahand. Chapter 10 applies the results in the first nine chapters to in-hand manipulation when the fingertips and the object maintain rolling contact.

The book presents a considerable number of examples with step-by-step solutions to balance ideas and facts. MATLAB™ has been used to produce some of the figures.

The targeted audience of this book include senior undergraduate students, graduate students, engineers, and researchers. The prerequisites for reading this book are linear algebra, calculus, and elementary knowledge in differential geometry. For reader's convenience, we have restricted our references to Martin M. Lipschutz's *Schaum's Outline of Differential Geometry* (McGraw-Hill Education Europe), Manfredo P. Do Carmo's *Differential Geometry of Curves and Surfaces* (Dover Publications Inc.), and Henri Cartan's *Differential Forms* (Dover Publications Inc.). Knowledge in the product-of-exponentials formula will be useful for understanding the contents in Chapters 8–10.

About the Authors

Lei Cui is a roboticist who focuses on developing theories of robotics and designs and develops novel robots and sensing technologies for real-world applications. He believes that both fundamental and applied research are indispensable to robotics: rigorous mathematical theories provide insight into applications, which in turn pose theoretical challenges. The goal of his work is to create sophisticated and intelligent robotic systems. He received his PhD degree from King's College London (UK), worked at Carnegie Mellon University (US), and is currently Senior Lecturer at Curtin University (AU).

Jian S. Dai is a kinematician and roboticist who developed screw theory and Lie groups in robotics Jacobian and for robotic manipulation that integrate Lie groups and finite screws in robotic trajectory planning and control, particularly for multi-fingered hand manipulation with geometry compatibility and elasticity restraint. The work develops a screw image space for grasping and for the metamorphic hand. He received a BEng in 1982 and an MSc in 1984 from Shanghai Jiao Tong University and received a PhD in Advanced Kinematics and Robotics from the University of Salford in the UK in

1993. With his continuing work, he received 2015 ASME Mechanisms and Robotics Award that is an honor to engineers and scientists who have made a lifelong contribution to the fundamental theory, design and applications of mechanisms and robotic systems as the 27th recipient since the award was established in 1974. Professor Dai has published over 500 peer-reviewed papers and 7 authored books. He is the Founder of the prestigious conference series *ASME/IEEE International Conference on Reconfigurable Mechanisms and Robots* (*ReMAR*) and has guided 32 PhD students.

Contents

Chapter 1

Introduction

1.1 A Brief History of Rolling–Sliding Contact and In-Hand Manipulation

An object rolls and slides on another is a common phenomenon in nature. Galileo rolled a ball down a slope, leading to the hypothesis: forces change motion.

When two general curved surfaces roll and slide on each other, the surface theory in differential geometry developed by Gauss (Gauss and Pesic, 2005) comes into play. Differential geometry is the study of curved spaces using the techniques of calculus. It is also the language used by Einstein to express general relativity, and so is an essential tool for astronomers and theoretical physicists.

Frame fields constitute a very useful tool for studying curves and surfaces (Cartan, 2002). Cartan's work was later developed by Chern (Chern *et al.*, 1999) and others (Cartan, 1996) into the moving-frame method, which is recognized as one of the basic techniques of research in modern differential geometry.

Rolling–sliding contact has been used in a variety of robotic applications: spherical robots, single-wheel robots, and multifingered robotic hands. Spherical robots, such as the BB-9E^{TM} in the Star Wars, are driven by a spherical shell for mobile surveillance and space exploration, and their electrical and electronic parts are sealed off inside the shell (Joshi *et al.*, 2010; Javadi and Mojabi, 2004; Li and Liu, 2011; Kim *et al.*, 2010). Dynamically balanced single-wheel robots can negotiate rough terrains at high speed without a steering device (Blažič, 2011; Dixon *et al.*, 2001, 2004; Soetanto *et al.*, 2003; Xu, 1999).

In-hand manipulation exemplifies the application of rolling contact, where a multifingered hand (Cui *et al.*, 2014; Dai *et al.*, 2009; Sarkar *et al.*, 1997; Ma and Dollar, 2013; Cole *et al.*, 1989) or a group of collaborative robotic arms (Kiss *et al.*, 2002; Sarkar *et al.*, 1997) employs rolling contact to steer an object to reach a goal configuration.

Rolling contact used for in-hand manipulation can reduce abrasion wear, simplify control, and enlarge reachable workspace. Further, curvatures of the surfaces in rolling can be taken advantage of to better control a multifingered hand. Hence, surface theory in differential geometry developed by Gauss and Pesic (2005), Cartan (2002) and Chern *et al.* (1999) has been applied to the kinematics of rolling contact.

Rolling contact engenders two geometric constraints. The first is the two normal vectors of the surfaces in contact coincide, and the second is the velocities of the two contact points are equal, i.e., the two contact trajectories are tangent to each other and have the same rolling rate. Sliding contact engenders only the first geometric constraint.

When maintaining rolling–sliding contact with another object, an object has five degrees-of-freedom (DOFs): three rotational and two translational. Taking away the sliding contact removes the two translational DOFs and only three rotational DOFs remain. Pure-rolling, which does not have the rotational (spin) motion about the normal direction, does not occur naturally: the two contact trajectories must have the same geodesic curvature. This was first discovered in the case of a convex surface rolling on a plane (Chitour *et al.*, 2005; Nomizu, 1978) and later proved rigorously when two general surfaces are concerned (Cui and Dai, 2010).

Rolling contact imposes nonholonomic constraints to mechanical systems, and most textbooks treat nonholonomic kinematics and dynamics briefly in one or two chapters — a disc rolling on a plane is normally used as a case study (Rosenberg, 1977; Neimark and Fufaev, 1972; Ardema, 2004; Bullo and Lewis, 2005).

The wide applications of tactile fingertips have steered the focus of the study of robotic hand from grasping to in-hand manipulation (Yousef *et al.*, 2011). Though re-grasping, which entails a sequence of pick and place (or throw) operations, is one technique (Sudsang and Phoka, 2003; Dafle *et al.*, 2014; Furukawa *et al.*, 2006), in-hand manipulation by rolling is more natural, typifying the dexterity of a human hand (Exner, 1989; Pehoski *et al.*, 1997). Hence, in-hand manipulation with rolling contact has garnered plenty of attention and been treated from a variety of aspects.

Kinematics of contact is greatly influenced by three mainstream formulations (Cai and Roth, 1986; Montana, 1988; Cui and Dai, 2010), which are closely related to different approaches in differential geometry. The first formulation is obtained by approximating surfaces locally via Taylor series; the second formulation considers local coordinates of parametric surfaces; and the third formulation is developed through the moving-frame method, gaining one important feature in modern differential geometry: coordinate-invariance.

Cai and Roth (1986) studied the kinematics of pure-rolling via time-based invariants (Roth, 2005), assuming that the surfaces as Monge patches (Carmo, 1976; Lipschutz, 1969) and expressing the surface mappings via Taylor series. Applying this work to in-hand manipulation, Kerr and Roth (1986) developed the movements of the fingers to achieve a desired motion of the object. This approach was also used by Harada *et al.* (2002) and Maekawa *et al.* (1997) where quadratic surfaces were used to approximate the local geometry of the fingertips and object.

Montana (1988, 1992, 1995) parameterized the surfaces as two-dimensional manifold embedded in 3D Euclidean space and derived a system of nonlinear ordinary differential equations in terms of the four surface parameters and the angle between the contact frames. This formulation was later extended to include acceleration terms that depends explicitly on the Christoffel symbols and their time derivatives (Sarkar *et al.*, 1996).

Cui and Dai (2010, 2015, 2018) and Cui *et al.* (2017) embedded the constraints into the moving-frame method along two contact trajectories to formulate the kinematics in terms of relative curvatures of the contact trajectories and further applied this formulation to in-hand manipulation. These works removed the restrictions of previous work (Neimark and Fufaev, 1972), where the moving-frame method was set up along the lines of curvatures. These works also clarified one misconception presented by Chen (1997): the magnitude of the spin angular velocity was arbitrary.

Kinematics of rolling–sliding contact is essential to control and path planning when two rigid surfaces are subject to these constraints. Some authors studied this issue from the point of view of control. Cole *et al.* (1989) were the first to develop the control of in-hand manipulation with rolling contact, where kinematics of pure-rolling was formulated in matrix form subject to four constraints. Sarkar *et al.* (1997) formulated the kinematics of pure-rolling in terms of the coefficients of the first and second fundamental forms to maintain rolling contact for in-hand manipulation. Remond *et al.* (2002) derived the kinematics of in-hand manipulation in six equations and four constraint equations, and acknowledged that the four constraint equations could not be derived in the general case and therefore proposed polar parameterization to approximate the surfaces. Agrachev and Sachkov (1999) used intrinsic form of pure-rolling constraints to discuss the controllability of two rigid objects subject to rolling contact.

Some works were concerned with motion planning of two rigid objects subject to rolling contact. Chelouah and Chitour (2003) and Chitour *et al.* (2005) studied the motion planning of a convex 2D Riemannian manifold rolling on a plane and established that the geometric curvatures of the two contact trajectories have to be equal to eliminate the spin motion. Tchon (2002) and Tchon and

Muszynski (1998) derived the Jacobian of a doubly nonholonomic mobile manipulator based on the concept of endogenous configuration space. Li and Canny (1990) and Marigo and Bicchi (2000) established the reachability of two surfaces under rolling contact based on the formulas (Montana, 1988) and provided algorithms for a sphere to reach a goal configuration on a plane.

Researchers have long recognized the importance of time-independent kinematics properties and instantaneous invariants (Veldkamp, 1967; Kirson and Yang, 1978; Roth, 2005, 2015; De Schutter, 2009) when dealing with the motion of a rigid body with respect to a reference frame. Geometric kinematics, which is formulated using arc length and curvatures, could reveal the characteristics of the trajectories elegantly and rigorously. These invariants combined with canonical systems had been the mainstream approach when studying instantaneous kinematics of rigid bodies.

A more general approach is to use tensors to express the kinematics of a rigid body, which has the advantage of being frame-invariant (or frame indifference) (Casey and Lam, 1986; Angeles, 1988; Legnani *et al.*, 1996; Cui, 2010). Curvatures are scalars, which are tensors of rank 0; differential one-forms are tensors of rank $(0, 1)$; and vectors are tensors of rank $(1, 0)$. These are geometric entities whose components are transformed in a certain way that passes from one reference frame to another, but they themselves do not change. In this sense, the kinematic formulations are coordinate invariant, which enables the formulations to be referenced in a convenient reference frame and later used in another frame.

The adjoint approach is a powerful tool to study geometry of curves and surfaces when two geometric entities are present. It is credited to be conceived by Cesaro (Sasaki, 1955), an Italian mathematician in the field of differential geometry. The idea is to study the geometric properties of a curve or a surface from the curve's adjoint curve or the surface's adjoint surface. This approach was first applied by Wang and Xiao (1993) to planar four-bar linkages, revealing the shapes of coupler curves by treating fixed centrodes as the adjoint curve. They later extended this approach to spatial linkages (Wang *et al.*, 1997a,b,c), where the line of striction of fixed axodes used as the adjoint curve, revealing the characteristics of the trajectories of points and lines.

The adjoint approach was further used to investigate the geometric properties of circular surfaces, which are swept out by moving a circle with its center following a spine curve (Cui *et al.*, 2009). The approach later facilitated the study of the forward and inverse kinematics of rolling–sliding contact between two rigid surfaces and demystified the causes of the spin motion (Cui and Dai, 2010, 2012, 2015, 2018; Cui *et al.*, 2017; Dai *et al.*, 2009). The results were recently adapted to solve the kinematics of contact of space-like surfaces in Lorentzian 3-space (AYDINALP *et al.*, 2018; Bayar *et al.*, 2018).

1.2 Outline of the Book

1.2.1 *Statement of the Problem*

In-hand manipulation involves manipulating an object by fingers without motion of the arm on which the hand is mounted. The contact points between a fingertip and the object could be moving or fixed. We focus on the first case, where the contact points generate two trajectories on the fingertip and on the object respectively.

Two types of contact in the first case can be discerned: rolling contact and sliding–rolling contact. A moving object maintaining rolling contact with a fixed one has three DOFs consisting of three angular motions with their directions passing the contact point. A moving object maintaining sliding–rolling contact has five DOFs including two translational DOFs in the common tangent plane and three angular DOFs with their directions passing through the contact point.

Previous literature used the term ‘point contact’ for all sliding–rolling, rolling, and fixed point contact. To avoid confusion, this book will use sliding–rolling contact both for general contact and for the contact where there exist relative sliding and rolling movements, and use rolling contact where there is only rolling but no sliding relative movement.

In-hand manipulation has to consider the geometric features of the fingertip and the object, which has attracted great interest recently, triggered mainly by the desire to conduct in-hand manipulations with tactile fingers. A number of kinematic formulations have been derived with various techniques. Recently researchers have started to realize that modern differential geometry could play an important role. One advantage offered by modern differential geometry is that it is able to give coordinate-independent kinematic formulations, which are highly desirable given many coordinate frames involved in a multifingered hand system. For example, a fixed reference frame is usually located at the palm and each joint has its own frame, and the two surfaces add another two frames.

A useful kinematic formulation of in-hand manipulation should consider computational efficiency, as well as the extent of ease it can be manipulated. Further, it should have the flexibility to admit a degree of coordinate independence, i.e., a given problem should not be confined to any specific choice of reference frames to carry out the kinematic analysis.

Motivated in part by these considerations, the product-of-exponentials (POE) formula has been widely used in analyzing kinematics of in-hand manipulation. However, kinematics of rolling or sliding–rolling contact is traditionally derived in terms of coordinates of surface patches, which nevertheless appear in the kinematic formulations. This to some extent compromises the initiative.

1.2.2 *Aims and Objectives*

This book focuses on the development of kinematics of rolling contact and sliding–rolling contact in a coordinate-independent form and applies the formulations to in-hand manipulation.

The following are the objectives of this book:

- Develop the kinematic formulations of spin–rolling from two contact trajectories when two rigid surfaces maintain rolling contact.
- Generate two contact trajectories from a given spin–rolling motion.

- Develop the kinematic formulations of sliding–spin–rolling from two contact trajectories when two rigid surfaces maintain sliding–rolling contact.
- Generate two contact trajectories from a given sliding–spin–rolling motion.
- Develop general kinematic equations of in-hand manipulation when the fingertips and the object maintain rolling contact.

1.3 Organization of the Book

This book is organized in 10 chapters with the following structure.

Chapter 1 presents the statement of the problem together with the aims and objectives of the book. The chapter describes the contents of the book, introduces background and current development, and puts the work conducted in the book with existing literature. It reviews the work on kinematics of rolling contact and sliding–rolling contact and analyzes the methodology used for the study.

Chapter 2 introduces the mathematical tools used in later chapters, including the moving-frame method, curvatures of curves and surfaces, the first and second fundamental forms, and the Darboux frame. The chapter reveals the relationship of the curvatures of the coordinate curves and those of the Darboux formulas, paving the way for the inverse kinematics of rolling contact.

Chapter 3 introduces the adjoint approach, which is an important tool to study the geometry of a curve or a surface from a companion curve or surface. The chapter demonstrates the adjoint approach in Roulettes and Bertrand curves and further presents the fixed-point conditions. Circular surfaces, which is an example of a surface adjoint to a curve, are also discussed.

Chapter 4 obtains a kinematic formulation of rolling contact when two contact trajectories are present. Embedding the constraints into the moving-frame method, the chapter uses the fixed-point conditions to derive the formulation in terms of rolling rate and surface curvatures along the contact trajectories. This formulation clarifies the factors that affect the spin motion: the rolling rate and the relevant geodesic curvature. The chapter then demonstrates that qualitative information about trajectories can be deduced from this formulation.

Chapter 5 extends the work in Chapter 4 for obtaining a kinematic formulation of rolling–sliding contact when two contact trajectories are present. This chapter reveals the geometric properties of the instantaneous imprint trajectory, proving that the angular velocity produced by the imprint trajectory is in accordance with the formulation presented in Chapter 3.

Chapter 6 generates contact trajectories from given spin–rolling velocity based on the kinematic formulations developed in Chapter 4. The chapter obtains three geometric entities: rolling rate, rolling direction, and compensatory spin rate, which can be used to realize the given instantaneous velocity.

Chapter 7 extends the work in Chapter 6 for generating contact trajectories from given sliding–spin–rolling velocity based on the kinematic formulations developed in Chapter 5. The chapter obtains five geometric entities: sliding rate, sliding direction, rolling rate, rolling direction, and compensatory spin rate, which can be used to realize the given instantaneous velocity.

Chapter 8 proposes a framework for the kinematics of in-hand manipulation in terms of the POE formula when the fingertips and the object maintain fixed-point contact. The chapter derives a singularity-free solution to the inverse kinematics of a robotic hand with an articulated palm — the Metahand.

Chapter 9 applies the concept of Gauss map to investigating the palm workspace and posture workspace of the Metahand, providing visually appealing representations of the workspaces.

Chapter 10 integrates the contents in the previous chapters and describes the kinematics of in-hand manipulation with rolling contact. The chapter examines the in-hand manipulation of a two-fingered planar hand and the the three-fingered Metahand.

Part I

The Moving-Frame Method and Adjoint Approach

Chapter 2

Curvatures of Curves and Surfaces via the Moving-Frame Method

In this chapter, we introduce the moving-frame method to study the differential properties of curves and surfaces. In particular, we obtain the infinitesimal displacement of the Frenet frame on a spatial curve in terms of curvature k and torsion τ and then the infinitesimal displacement of the Darboux frame on a surface curve in terms of geodesic curvature k_g, normal curvature k_n, and geodesic torsion τ_g.

2.1 The Moving-Frame Method

We consider a transformation T of the linear-affine group in Euclidean n-space $\mathbb{R}^n$. An affine frame $(\mathbf{M}, \mathbf{e}_1, \ldots, \mathbf{e}_n)$ consists of the point $\mathbf{M}$ that is the image of the origin $\mathbf{O} \in \mathbb{R}^n$ and the vectors $\mathbf{e}_1, \ldots, \mathbf{e}_n$ that are respectively the images of the canonical base vectors $(1, 0, 0, \ldots, 0), \ldots, (0, 0, 0, \ldots, 1)$ of $\mathbb{R}^n$ under the transformation T subject to the condition

$$\det(\mathbf{e}_1, \ldots, \mathbf{e}_n) \neq 0$$

The vectors $\mathbf{e}_1, \ldots, \mathbf{e}_n$ form a base of $\mathbb{R}^n$.

We call the affine frame $(\mathbf{M}, \mathbf{e}_1, \ldots, \mathbf{e}_n)$ that depends on one or more real parameters a moving frame. We assume that $\mathbf{M}, \mathbf{e}_1, \ldots, \mathbf{e}_n$ are differentiable vector-valued functions of an open set $U \subset \mathbb{R}^n$ to $\mathbb{R}^n$. It follows that differentials $d\mathbf{M}$ and $d\mathbf{e}_i$ can be written as linear combinations of the base vectors $\mathbf{e}_1, \ldots, \mathbf{e}_n$ as

$$d\mathbf{M} = \sum_{i=1}^{n} \omega_i \mathbf{e}_i \tag{2.1}$$

$$d\mathbf{e}_i = \sum_{j=1}^{n} \omega_{ij} \mathbf{e}_j \tag{2.2}$$

where ω_i and ω_{ij} are scalar-valued differential forms of degree 1 in U, and these differential forms define the infinitesimal displacement of the moving frame $(\mathbf{M}, \mathbf{e}_1, \ldots, \mathbf{e}_n)$.

It is advantageous to investigate the geometry of curves and surfaces through the right-handed orthonormal frame, where the base vectors $\mathbf{e}_1, \ldots, \mathbf{e}_n$ are of unit length and mutually orthogonal, and $\det(\mathbf{e}_1, \ldots, \mathbf{e}_n) = 1$. It follows that

$$\mathbf{e}_i \cdot \mathbf{e}_j = \begin{cases} 1 & i = j \\ 0 & i \neq j \end{cases}$$

Differentiating the above equation gives

$$d\mathbf{e}_i \cdot \mathbf{e}_j + \mathbf{e}_i \cdot d\mathbf{e}_j = 0$$

Substituting $d\mathbf{e}_i$ and $d\mathbf{e}_j$ with the linear combinations in Eq. (2.2) yields

$$\omega_{ij} = -\omega_{ji} \tag{2.3}$$

This implies $\omega_{ii} = 0$ in particular.

We examine the relationship between the one-forms $\omega_i, i = 1, 2, 3$ and ω_{ij}, $i, j = 1, 2, 3$ of the right-handed orthonormal frame $(\mathbf{M}, \mathbf{e}_1, \ldots, \mathbf{e}_n)$.

For any differentiable function f, we have $d^2 f = 0$. It follows that

$$d^2\mathbf{M} = 0, \quad d^2\mathbf{e}_i = 0$$

It follows from Eq. (2.1) that

$$0 = d^2\mathbf{M} = d\left(\sum_{i=1}^{n} \omega_i \mathbf{e}_i\right)$$

$$= \sum_{i=1}^{n} (d\omega_i \mathbf{e}_i - \omega_i \wedge d\mathbf{e}_i) = \sum_{i=1}^{n} d\omega_i \mathbf{e}_i - \sum_{i=1}^{n} \omega_j \wedge d\mathbf{e}_i$$

$$= \sum_{i=1}^{n} d\omega_i \mathbf{e}_i - \sum_{i=1}^{n} \omega_i \wedge (\sum_{j=1}^{n} \omega_{ij} \mathbf{e}_j) = \sum_{i=1}^{n} \left(d\omega_i - \sum_{j=1}^{n} \omega_j \wedge \omega ij\right) \mathbf{e}_i$$

Since the vectors $\mathbf{e}_1$–$\mathbf{e}_n$ form an orthonormal basis, it follows that

$$d\omega_i = \sum_{j=1}^{n} \omega_j \wedge \omega ij$$

Similarly, $d^2\mathbf{e}_i = 0$ gives the following relationship:

$$d\omega_{ij} = \sum_{k=1}^{n} \omega_{ik} \wedge \omega_{kj}$$

2.2 The Frenet Frame of an Oriented Curve

A parameterized differentiable curve L is defined as a differentiable map $\mathbf{x} : (a, b) \to \mathbb{R}^3$, where (a, b) is an open interval of the real line $\mathbb{R}$. A curve can be represented as

$$\mathbf{x}(t) = x_1(t)\mathbf{i} + x_2(t)\mathbf{j} + x_3(t)\mathbf{k} \tag{2.4}$$

where $t \in (a, b)$ and $x_1(t)$–$x_3(t)$ are differentiable, and $\mathbf{i} = (1, 0, 0)$, $\mathbf{j} = (0, 1, 0)$, and $\mathbf{k} = (0, 0, 1)$ are the canonical base vectors of $\mathbb{R}^3$.

A differentiable curve L is oriented if a field of unit tangent vectors to L is differentiable. We assume the curves in this book are sufficiently differentiable and oriented.

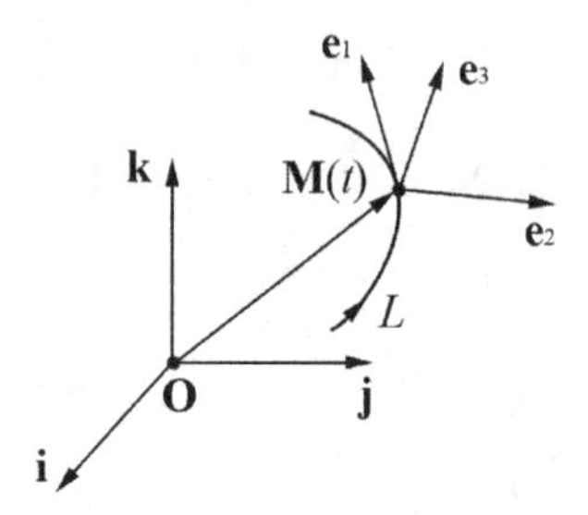

Fig. 2.1: The Frenet frame $(\mathbf{M}(t),\ \mathbf{e}_1(t),\ \mathbf{e}_2(t),\ \mathbf{e}_3(t))$ along the oriented curve L

We associate the following right-handed orthonormal frame to each point M of the curve L: $\mathbf{e}_1(t)$ is the unit tangent vector to the curve L oriented in the direction of increasing t, $\mathbf{e}_2(t)$ is the unit vector along the direction of $d\mathbf{e}_1(t)/dt$, and $\mathbf{e}_3(t) = \mathbf{e}_1(t) \times \mathbf{e}_2(t)$ so that $(\mathbf{M}(t),\ \mathbf{e}_1(t),\ \mathbf{e}_2(t),\ \mathbf{e}_3(t))$ is a right-handed orthonormal frame. Here, we use $\mathbf{M}(t)$ to represent the vector from the origin O to the point M, as shown in Fig. 2.1.

Definition 2.1. The Frenet Frame. The moving frame $(\mathbf{M},\ \mathbf{e}_1,\ \mathbf{e}_2,\ \mathbf{e}_3)$ defined in the aforementioned way along an oriented curve L is called the Frenet frame.

Since the vector $\mathbf{e}_1$ is tangent to L, it follows that

$$d\mathbf{M} = \omega_1 \mathbf{e}_1 \tag{2.5}$$

The differential forms ω_2 and ω_3 in Eq. (2.1) are 0 subsequently. The vector $\mathbf{e}_2$ is chosen to be the unit vector along the direction of $d\mathbf{e}_1(t)$. It follows that

$$d\mathbf{e}_1 = \omega_{12} \mathbf{e}_2 \tag{2.6}$$

The differential forms ω_{11}, ω_{13} in Eq. (2.2) are 0 subsequently. The differential $d\mathbf{e}_2 = \omega_{21}\mathbf{e}_1 + \omega_{22}\mathbf{e}_2 + \omega_{23}\mathbf{e}_3$, but $\omega_{21} = -\omega_{12}$ and $\omega_{22} = 0$ from Eq. (2.3). It follows that

$$d\mathbf{e}_2 = -\omega_{12}\mathbf{e}_1 + \omega_{23}\mathbf{e}_3 \tag{2.7}$$

Similarly, we have

$$d\mathbf{e}_3 = -\omega_{23}\mathbf{e}_2 \tag{2.8}$$

where $\omega_{31} = -\omega_{13} = 0$, $\omega_{32} = -\omega_{23}$, and $\omega_{33} = 0$.

Equations (2.5)–(2.8) give the infinitesimal displacement of the moving frame $(\mathbf{M}, \mathbf{e}_1, \mathbf{e}_2, \mathbf{e}_3)$, which can be written in matrix form as

$$d\mathbf{M} = \omega_1\mathbf{e}_1$$

$$d\begin{bmatrix}\mathbf{e}_1\\ \mathbf{e}_2\\ \mathbf{e}_3\end{bmatrix} = \begin{bmatrix} 0 & \omega_{12} & 0\\ -\omega_{12} & 0 & \omega_{23}\\ 0 & -\omega_{23} & 0\end{bmatrix}\begin{bmatrix}\mathbf{e}_1\\ \mathbf{e}_2\\ \mathbf{e}_3\end{bmatrix} \tag{2.9}$$

Note that the differential forms ω_1, ω_{12}, and ω_{23} are scalar-valued functions of variable t. We examine the geometric meaning of each differential form in the subsequent sections.

2.3 The Arc Length and Curvature of a Curve

Let s denote the arc length between $\mathbf{M}(t_0)$ and $\mathbf{M}(t)$ of the curve L $(t \geq t_0)$. It follows that

$$s = s(t) = \int_{t_0}^{t} \left|\frac{d\mathbf{M}}{dt}\right| dt$$

and

$$\frac{ds}{dt} = \frac{d}{dt}\left(\int_{t_0}^{t} \left|\frac{d\mathbf{M}}{dt}\right| dt\right) = \left|\frac{d\mathbf{M}}{dt}\right| \tag{2.10}$$

Note the curve L can be parameterized by the arc length s since Eq. (2.10) has a continuous non-vanishing derivative. We call s the *natural parameter*, and the parameterization of the curve L by its natural parameter s a *natural representation.*

The curve L parameterized by s is of *unit speed*, which is given by

$$\frac{d\mathbf{M}}{ds} = \frac{d\mathbf{M}/dt}{ds/dt} = \frac{d\mathbf{M}/dt}{|d\mathbf{M}/dt|} = \mathbf{e}_1 \tag{2.11}$$

Equation (2.5) gives $d\mathbf{M}/\omega_1 = \mathbf{e}_1$. It thus follows from Eqs. (2.5) and (2.11) that

$$\omega_1 = ds \tag{2.12}$$

This gives the geometric meaning of the differential form ω_1.

The straight line passing the point M of L and along the direction of $\mathbf{e}_1$ is called the *tangent line* to L at the point M, which is given by

$$\mathbf{x} = \mathbf{M} + t\mathbf{e}_1, \quad t \in \mathbb{R}$$

The plane through the point M perpendicular to the tangent line is called the normal plane, as shown in Fig. 2.2, which is given by

$$(\mathbf{x} - \mathbf{M}) \cdot \mathbf{e}_1 = 0$$

The differentiation of the unit tangent vector $\mathbf{e}_1$ w.r.t. the arc length s, $d\mathbf{e}_1/ds$, is called the *curvature vector*, which is denoted by $\mathbf{k}$. Note that the curvature vector $\mathbf{k}$ is perpendicular to the vector $\mathbf{e}_1$ and lies in the normal plane, since $\mathbf{e}_1$ is of unit length, as shown in Fig. 2.3.

The *curvature* k is the rate of change of the tangent line w.r.t. the arc length s, which is given by

$$\frac{d\mathbf{e}_1}{ds} = \mathbf{k} = k\mathbf{e}_2 \tag{2.13}$$

The unit vector $\mathbf{e}_2$ is parallel to the curvature vector $\mathbf{k}$, and is called the *principal normal vector* to L at M. The reciprocal of the

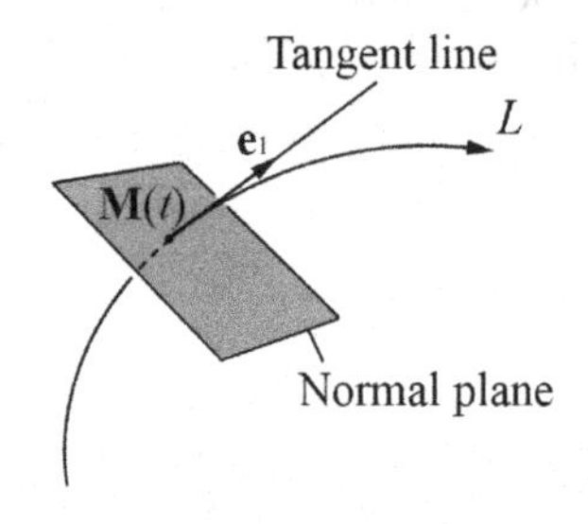

Fig. 2.2: The tangent line and the normal plane at the point M

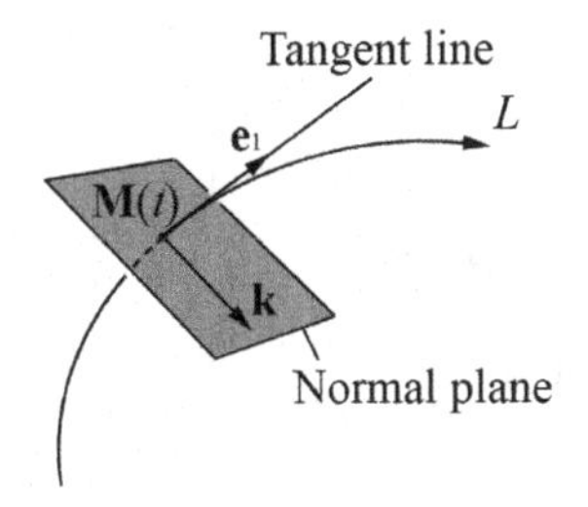

Fig. 2.3: The curvature vector at the point M

curvature is called the *radius of curvature*, which is given by

$$\rho = \frac{1}{k} \tag{2.14}$$

Remark 2.1. Geometrically, the curvature k measures how sharply the curve bends. A large curvature (or small radius of curvature) means the curve bends more sharply; a small curvature (or large radius of curvature) means the curve bends less sharply. In particular, the curvature of a straight line is 0.

Remark 2.2. The curvature defined in Eq. (2.13) is called the *signed curvature*, the value of which depends on the direction of the principal normal vector $\mathbf{e}_2$.

Equations (2.6) and (2.12) give the curvature vector $\mathbf{k}$ in terms of the differential forms as

$$\mathbf{k} = \frac{d\mathbf{e}_1}{ds} = \frac{\omega_{12}\mathbf{e}_2}{\omega_1} \tag{2.15}$$

It follows that

$$k = \frac{\omega_{12}}{\omega_1} \tag{2.16}$$

This gives the geometric meaning of the differential form ω_{12}.

The straight line passing through the point M of L and along the direction of $\mathbf{e}_2$ is called the *principal normal line*, which is given by

$$\mathbf{x} = \mathbf{M} + t\mathbf{e}_2, \quad t \in \mathbb{R} \tag{2.17}$$

The plane spanned by the unit tangent vector $\mathbf{e}_1$ and the principal normal vector $\mathbf{e}_2$ is called the *osculating plane*, as shown in Fig. 2.4.

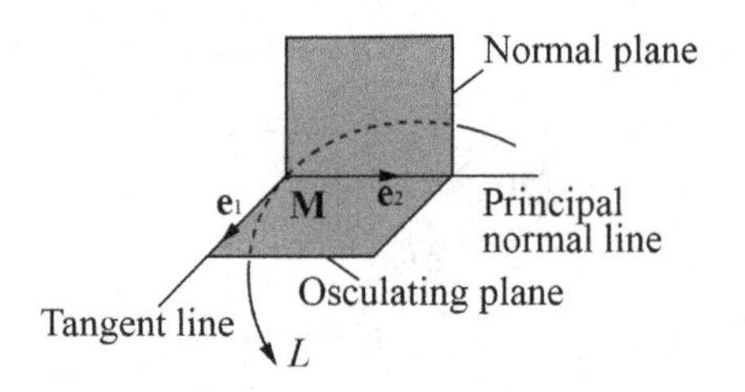

Fig. 2.4: The principal normal line and osculating plane

2.4 The Torsion of a Curve

The unit vector $\mathbf{e}_3$ that is perpendicular to the tangent vector $\mathbf{e}_1$ and principal normal vector $\mathbf{e}_2$ is called the *binormal vector* to L at the point M.

The straight line passing through the point M along the direction of $\mathbf{e}_3$ is called the *binormal line*, which is given by

$$\mathbf{x} = \mathbf{M} + t\mathbf{e}_3, \quad t \in \mathbb{R} \tag{2.18}$$

The plane spanned by the unit tangent vector $\mathbf{e}_1$ and binormal vector $\mathbf{e}_3$ is called the *rectifying plane* at the point M, which is given by

$$(\mathbf{x} - \mathbf{M}) \cdot \mathbf{e}_2 = 0 \tag{2.19}$$

We have thus defined three characteristic lines — the tangent line, principal line, and binormal line — and three characteristic planes — the normal plane, osculating plane, and rectifying plane — at the point M, as shown in Fig. 2.5.

The derivative of $\mathbf{e}_2$ w.r.t. the arc length s is perpendicular to $\mathbf{e}_2$ since $\mathbf{e}_2$ is of unit length. We thus can project $d\mathbf{e}_2/ds$ onto the $\mathbf{e}_1$ and $\mathbf{e}_3$-axes. It follows that

$$\frac{d\mathbf{e}_2}{ds} = \alpha\mathbf{e}_1 + \tau\mathbf{e}_3 \tag{2.20}$$

Here, τ is called the *torsion* of L at the point M.

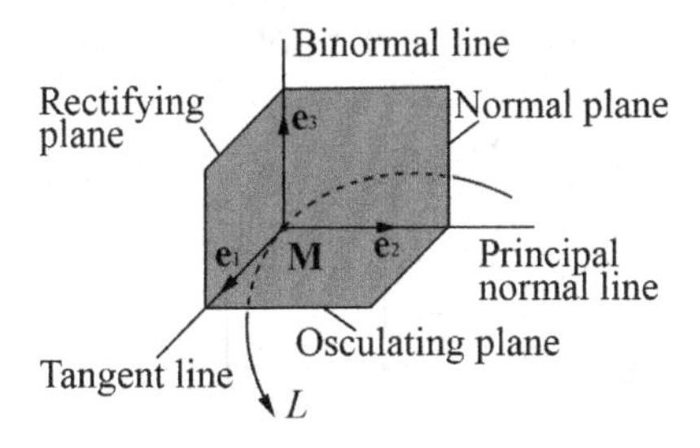

Fig. 2.5: The three characteristic lines and three planes

Differentiating the binormal vector $\mathbf{e}_3$ w.r.t. the arc length s yields

$$\begin{aligned}\frac{d\mathbf{e}_3}{ds} &= \frac{d(\mathbf{e}_1 \times \mathbf{e}_2)}{ds} = \frac{d\mathbf{e}_1}{ds} \times \mathbf{e}_2 + \mathbf{e}_1 \times \frac{d\mathbf{e}_2}{ds} \\ &= k\mathbf{e}_2 \times \mathbf{e}_2 + \mathbf{e}_1 \times (\alpha\mathbf{e}_1 + \tau\mathbf{e}_3) = -\tau\mathbf{e}_2 \end{aligned} \tag{2.21}$$

The value of α can be straightforwardly determined as $-k$ because of the antisymmetry of the matrix in Eq. (2.3). Equations (2.8) and (2.21) give the torsion in terms of the differential forms as

$$\tau = \frac{\omega_{23}}{\omega_1} \tag{2.22}$$

Remark 2.3. Geometrically, the torsion measures the rate of change of the curve pulling away from the osculating plane. A large torsion means the curve pulls away from the osculating plane quickly; a small torsion means the curve pulls away from the osculating plane slowly. In particular, the torsion of a planar curve is 0.

Definition 2.2. The Frenet formulas. With the introduction of the arc length s, the curvature k, and the torsion τ, the Frenet formulas are defined as

$$\begin{aligned}&\frac{d\mathbf{M}}{ds} = \mathbf{e}_1 \\ &\frac{d}{ds}\begin{bmatrix}\mathbf{e}_1 \\ \mathbf{e}_2 \\ \mathbf{e}_3\end{bmatrix} = \begin{bmatrix}0 & k & 0 \\ -k & 0 & \tau \\ 0 & -\tau & 0\end{bmatrix}\begin{bmatrix}\mathbf{e}_1 \\ \mathbf{e}_2 \\ \mathbf{e}_3\end{bmatrix}\end{aligned} \tag{2.23}$$

In particular, the Frenet formulas of a planar curve are

$$\begin{aligned} \frac{d\mathbf{M}}{ds} &= \mathbf{e}_1 \\ \frac{d}{ds}\begin{bmatrix}\mathbf{e}_1\\ \mathbf{e}_2\end{bmatrix} &= \begin{bmatrix}0 & k\\ -k & 0\end{bmatrix}\begin{bmatrix}\mathbf{e}_1\\ \mathbf{e}_2\end{bmatrix} \end{aligned} \tag{2.24}$$

Remark 2.4. It can be proved that the curvature $k(s)$ and the torsion $\tau(s)$ as functions of the natural parameter s completely define a curve except for its position in space.

Example 2.1. A circle of radius R is centered at the point (a, b) in a plane, as shown in Fig. 2.6. Obtain the Frenet formulas of the circle.

Solution:

The circle L can be parameterized as

$$\mathbf{x}(t) = (R\cos(t) + a)\mathbf{i} + (R\sin(t) + b)\mathbf{j}$$

Let M denote a point on the circle L. From now on, we use 2-tuples to represent row vectors with respect to frame $(\mathbf{O}, \mathbf{i}, \mathbf{j})$.

Differentiating $\mathbf{M}$ yields

$$d\mathbf{M} = \begin{bmatrix}-R\sin(t)dt, & R\cos(t)dt\end{bmatrix}$$

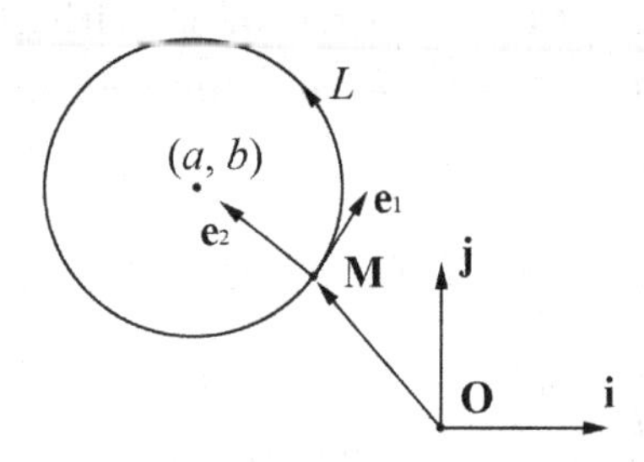

Fig. 2.6: The Frenet frame on the planar circle L

Since $d\mathbf{M} = \omega_1 \mathbf{e}_1$, the differential form ω_1 and the unit tangent vector $\mathbf{e}_1$ can be obtained respectively as

$$\omega_1 = Rdt$$

$$\mathbf{e}_1 = \begin{bmatrix} -\sin(t) & \cos(t) \end{bmatrix}$$

Differentiating the tangent vector $\mathbf{e}_1$ yields

$$d\mathbf{e}_1 = [-\cos(t)dt - \sin(t)dt] = dt[-\cos(t) - \sin(t)]$$

Since $d\mathbf{e}_1 = \omega_{12}\mathbf{e}_2$, it follows that

$$\omega_{12} = dt$$

$$\mathbf{e}_2 = \begin{bmatrix} -\cos(t) & -\sin(t) \end{bmatrix}$$

The curvature k can be obtained from Eq. (2.16) as

$$k = \frac{\omega_{12}}{\omega_1} = \frac{1}{R}$$

The Frenet formulas can be obtained from Eq. (2.24) as

$$\frac{d\mathbf{M}}{ds} = \mathbf{e}_1$$

$$\frac{d}{ds}\begin{bmatrix} \mathbf{e}_1 \\ \mathbf{e}_2 \end{bmatrix} = \begin{bmatrix} 0 & \frac{1}{R} \\ -\frac{1}{R} & 0 \end{bmatrix} \begin{bmatrix} \mathbf{e}_1 \\ \mathbf{e}_2 \end{bmatrix}$$ □

Remark 2.5. If we reverse the direction of $\mathbf{e}_2$, which points outward of the circle, the differential form ω_{12} is equal to $-dt$ and $\mathbf{e}_2$ is equal to $[\cos(t)\ \sin(t)]$. The curvature k is then equal to $-\frac{1}{R}$.

Example 2.2. Obtain the Frenet formulas of the helix, as shown in Fig. 2.7, which is given by the following equation:

$$\mathbf{x}(t) = \cos(t)\mathbf{i} + \sin(t)\mathbf{j} + t\mathbf{k}$$

Solution:

Let M denote a point on the curve helix. From now on, we use 3-tuples to represent row vectors with respect to the frame $(\mathbf{O}, \mathbf{i}, \mathbf{j}, \mathbf{k})$.

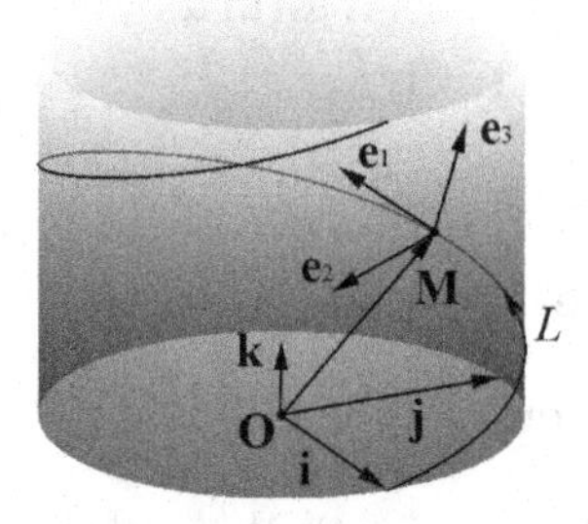

Fig. 2.7: The Frenet frame on helix L on a cylinder

It follows that

$$d\mathbf{M} = [-\sin(t)dt\ \cos(t)dt\ dt] = \sqrt{2}dt\left[-\frac{\sqrt{2}\sin(t)}{2}\ \frac{\sqrt{2}\cos(t)}{2}\ \frac{\sqrt{2}}{2}\right]$$

Since $d\mathbf{M} = \omega_1\mathbf{e}_1$, as shown in Eq. (2.5), the differential form ω_1 and the unit tangent vector $\mathbf{e}_1$ can be obtained respectively as

$$\omega_1 = \sqrt{2}dt$$

$$\mathbf{e}_1 = \left[-\frac{\sqrt{2}\sin(t)}{2}\ \frac{\sqrt{2}\cos(t)}{2}\ \frac{\sqrt{2}}{2}\right]$$

Differentiating the tangent vector $\mathbf{e}_1$ yields

$$d\mathbf{e}_1 = \left[-\frac{\sqrt{2}\cos(t)dt}{2}\ -\frac{\sqrt{2}\sin(t)dt}{2}\ 0\right] = \frac{\sqrt{2}dt}{2}[-\cos(t)\ -\sin(t)\ 0]$$

Since $d\mathbf{e}_1 = \omega_{12}\mathbf{e}_2$, as in Eq. (2.6), it follows that

$$\omega_{12} = \frac{\sqrt{2}dt}{2}$$

$$\mathbf{e}_2 = [-\cos(t)\ -\sin(t)\ 0]$$

The right-hand rule gives $\mathbf{e}_3$ as

$$\mathbf{e}_3 = \mathbf{e}_1 \times \mathbf{e}_2 = \left[\frac{\sqrt{2}\sin(t)}{2}\ -\frac{\sqrt{2}\cos(t)}{2}\ \frac{\sqrt{2}}{2}\right]$$

With $d\mathbf{e}_2 = [\sin(t)dt, \; -\cos(t)dt, \; 0]$, the differential form ω_{23} can be obtained from Eq. (2.7) as

$$\omega_{23} = d\mathbf{e}_2 \cdot \mathbf{e}_3 = \frac{\sqrt{2}dt}{2}$$

The curvature k and the torsion τ can be obtained from Eq. (2.11) as

$$k = \frac{\omega_{12}}{\omega_1} = \frac{\sqrt{2}dt/2}{\sqrt{2}dt} = \frac{1}{2}$$

$$\tau = \frac{\omega_{23}}{\omega_1} = \frac{\sqrt{2}dt/2}{\sqrt{2}dt} = \frac{1}{2}$$

The Frenet formulas can thus be obtained from Eq. (2.12) as

$$\frac{d\mathbf{M}}{ds} = \mathbf{e}_1$$

$$\frac{d}{ds}\begin{bmatrix}\mathbf{e}_1\\ \mathbf{e}_2\\ \mathbf{e}_3\end{bmatrix} = \begin{bmatrix}0 & 1/2 & 0\\ -1/2 & 0 & 1/2\\ 0 & -1/2 & 0\end{bmatrix}\begin{bmatrix}\mathbf{e}_1\\ \mathbf{e}_2\\ \mathbf{e}_3\end{bmatrix}$$

□

Remark 2.6. If we reverse the direction of $\mathbf{e}_2$, then $\mathbf{e}_2$ is equal to $[\cos(t)\; \sin(t)\; 0]$ and ω_{12} is equal to $-\sqrt{2}dt$. We will have $k = -1/2$ and $\tau = 1/2$.

Example 2.3. Obtain the Frenet formulas of the curve L, as in Fig. 2.8, which is given by the following equation.

$$\mathbf{x}(t) = [3t - t^3\; 3t^2\; 3t + t^3]$$

Solution:

Let M denote a point on the curve L. It follows that

$$d\mathbf{M} = dt[-3t^2{+}3\; 6t\; 3t^2{+}3] = 3\sqrt{2}(t^2{+}1)dt\left[-\frac{\sqrt{2}(t^2-1)}{2(t^2+1)}\; \frac{\sqrt{2}t}{t^2+1}\; \frac{\sqrt{2}}{2}\right]$$

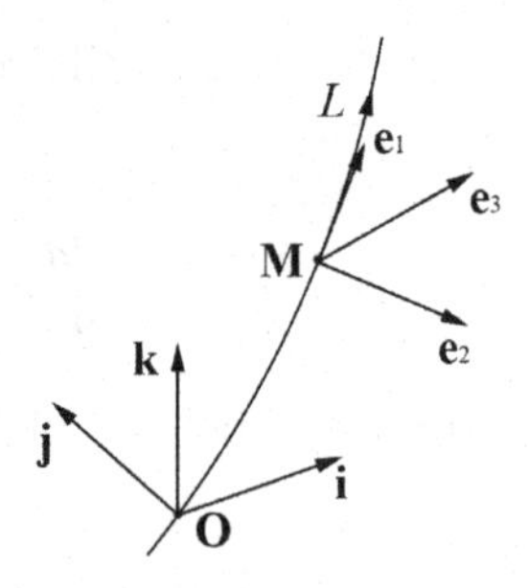

Fig. 2.8: The Frenet frame on the curve L

Since $d\mathbf{M} = \omega_1 \mathbf{e}_1$, it follows that

$$\omega_1 = 3\sqrt{2}(t^2+1)dt$$

$$\mathbf{e}_1 = \begin{bmatrix} -\frac{\sqrt{2}(t^2-1)}{2(t^2+1)} & \frac{\sqrt{2}t}{t^2+1} & \frac{\sqrt{2}}{2} \end{bmatrix}$$

Differentiating $\mathbf{e}_1$ yields

$$d\mathbf{e}_1 = \begin{bmatrix} -\frac{2\sqrt{2}tdt}{(t^2+1)^2} & -\frac{\sqrt{2}(t^2-1)dt}{(t^2+1)^2} & 0 \end{bmatrix} = -\frac{\sqrt{2}dt}{t^2+1} \begin{bmatrix} \frac{2t}{t^2+1} & \frac{t^2-1}{t^2+1} & 0 \end{bmatrix}$$

Since $d\mathbf{e}_1 = \omega_{12}\mathbf{e}_2$, it follows that

$$\omega_{12} = -\frac{\sqrt{2}dt}{t^2+1}$$

$$\mathbf{e}_2 = \begin{bmatrix} \frac{2t}{t^2+1} & \frac{t^2-1}{t^2+1} & 0 \end{bmatrix}$$

The right-hand rule gives the unit vector $\mathbf{e}_3$ as

$$\mathbf{e}_3 = \mathbf{e}_1 \times \mathbf{e}_2 = \begin{bmatrix} -\frac{\sqrt{2}(t^2-1)}{2(t^2+1)} & \frac{\sqrt{2}t}{t^2+1} & -\frac{\sqrt{2}}{2} \end{bmatrix}$$

The differential form ω_{23} can be obtained from Eq. (2.7) as

$$\omega_{23} = d\mathbf{e}_2 \cdot \mathbf{e}_3 = \frac{\sqrt{2}dt}{t^2+1}$$

The curvature k and the torsion τ can be obtained from Eqs. (2.16) and (2.22) respectively as

$$k = -\frac{1}{3(t^2+1)^2}$$
$$\tau = \frac{1}{3(t^2+1)^2}$$

We have obtained all the items of the Frenet formulas: $\mathbf{e}_1$, $\mathbf{e}_2$, $\mathbf{e}_3$, k, and τ. □

2.5 Differential Geometry of Surfaces

2.5.1 *Parameterized Regular Surfaces*

Definition 2.3. Regular surface. A regular surface is a subset $S \subset \mathbb{R}^3$ such that for any point M in S there exist an open subspace $U \subset \mathbb{R}^2$ and a map $\mathbf{x} : U \to S$ of the form:

$$\mathbf{x}(u,v) = \left[x_1(u,v),\, x_2(u,v),\, x_3(u,v)\right] \quad (u,v) \in U$$

where the map $\mathbf{x}$ has the following properties:

(1) $\mathbf{x}$ is differentiable. This means the components x_1, x_2, and x_3 of the map $\mathbf{x}$ have continuous partial derivatives of sufficiently high order.
(2) For any Q in U, the differential $d\mathbf{x}_Q : \mathbb{R}^2 \to \mathbb{R}^3$ is one-to-one.
(3) $\mathbf{x}$ is homeomorphism. This means there exists an open subspace $V \subset S$ that contains M such that $\mathbf{x}$ has an inverse $\mathbf{x}^{-1} : V \cap S \to U$ that is continuous.

Note the second property guarantees that a field of normal vectors exists on the surface S. A regular surface is called an oriented surface if the field of normal vectors is differentiable. The majority of surfaces involved in robotics are regular oriented surfaces, thus we assume such in this book.

The variables u and v are called parameters of the surface, and the first- and second-order partial derivatives are respectively:

$$\mathbf{x}_u = \frac{\partial \mathbf{x}}{\partial u} = \left[\frac{\partial x_1}{\partial u}, \frac{\partial x_2}{\partial u}, \frac{\partial x_3}{\partial u}\right]$$

$$\mathbf{x}_v = \frac{\partial \mathbf{x}}{\partial v} = \left[\frac{\partial x_1}{\partial v}, \frac{\partial x_2}{\partial v}, \frac{\partial x_3}{\partial v}\right]$$

$$\mathbf{x}_{uu} = \frac{\partial^2 \mathbf{x}}{\partial u^2} = \left[\frac{\partial^2 x_1}{\partial u^2}, \frac{\partial^2 x_2}{\partial u^2}, \frac{\partial^2 x_3}{\partial u^2}\right]$$

$$\mathbf{x}_{vv} = \frac{\partial^2 \mathbf{x}}{\partial v^2} = \left[\frac{\partial^2 x_1}{\partial v^2}, \frac{\partial^2 x_2}{\partial v^2}, \frac{\partial^2 x_3}{\partial v^2}\right]$$

$$\mathbf{x}_{uv} = \frac{\partial^2 \mathbf{x}}{\partial u \partial v} = \left[\frac{\partial^2 x_1}{\partial u \partial v}, \frac{\partial^2 x_2}{\partial u \partial v}, \frac{\partial^2 x_3}{\partial u \partial v}\right] = \mathbf{x}_{vu}$$

The pair $(U, \mathbf{x})$ is called a chart (a local coordinate system) around M.

An oriented differentiable curve L traced on the surface S can be parameterized as

$$\mathbf{x}(u(t), v(t)) = [x_1(u(t), v(t)),\ x_2(u(t), v(t)),\ x_3(u(t), v(t))]\ (u, v) \in U$$

In particular, the coordinate curves are defined as follows.

Definition 2.4. Coordinate curves. A u-curve $\mathbf{x} = \mathbf{x}(u, v_0)$ with u as the parameter is the image of the coordinate line $v = v_0$ in U. Similarly, a v-curve $\mathbf{x} = \mathbf{x}(u_0, v)$ with v as the parameter is the image of the coordinate line $u = u_0$ in U, as shown in Fig. 2.9. They are called the coordinate curves of the surface $\mathbf{x} = \mathbf{x}(u, v)$.

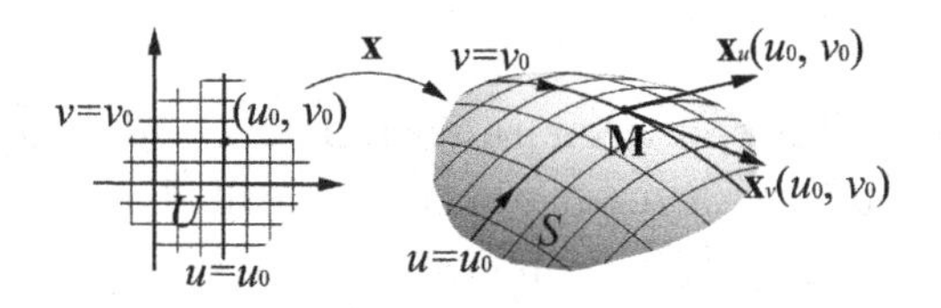

Fig. 2.9: The coordinate curves

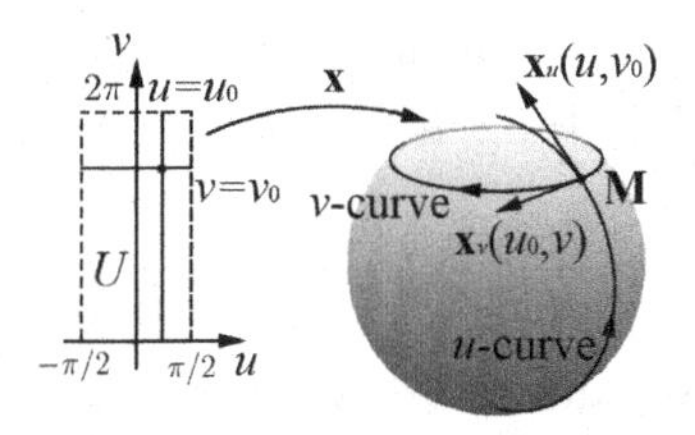

Fig. 2.10: The u-curve and v-curve on a sphere of radius r

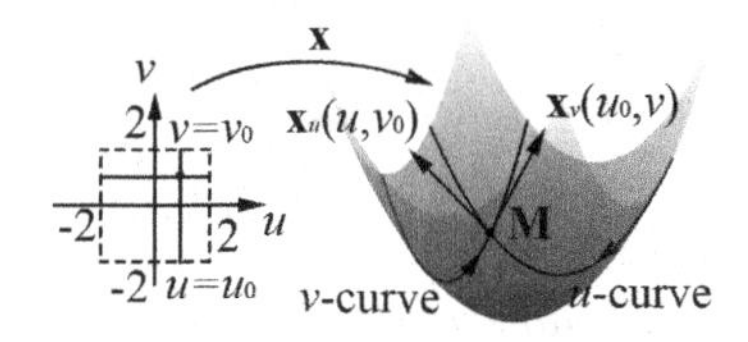

Fig. 2.11: The u-curve and v-curve on a paraboloid of revolution

Example 2.4. A coordinate patch of a sphere of radius R can be parameterized as

$$\mathbf{x}(u,v) = R[\cos(u)\cos(v), -\cos(u)\sin(v), \sin(u)]$$
$$u \in (-\pi/2, \pi/2), v \in [0, 2\pi)$$

The u-curve where $v_0 = 3\pi/2$ and the v-curve where $u_0 = \pi/4$ are illustrated in Fig. 2.10.

Example 2.5. A coordinate patch of a paraboloid of revolution is given as

$$\mathbf{x}(u,v) = \left[u,\ v,\ u^2 + v^2\right]$$
$$u \in (-2, 2), v \in (-2, 2)$$

The u-curve where $v_0 = 1$ and the v-curve where $u_0 = 1$ are illustrated in Fig. 2.11.

2.5.2 *The First Fundamental Form*

The first fundamental form is concerned with measurements on a surface: lengths of curves, angles of tangent vectors, and areas of regions.

A curve traced on the surface $\mathbf{x}(u, v)$ can be parameterized as $\mathbf{x}(t) = \mathbf{x}(u(t), v(t))$. The tangent vector at $\mathbf{x}$ can be obtained as

$$\frac{d\mathbf{x}}{dt} = \mathbf{x}_u \frac{du}{dt} + \mathbf{x}_v \frac{dv}{dt}$$

The derivatives du/dt and dv/dt are the coefficients of the tangent vector expressed w.r.t. the base $\mathbf{x}_u$ and $\mathbf{x}_v$. This equation in differential notation is $d\mathbf{x} = \mathbf{x}_u du + \mathbf{x}_v dv$.

We consider the square of the length of the tangent vector $d\mathbf{x}$:

$$\begin{aligned} I &= d\mathbf{x} \cdot d\mathbf{x} = (\mathbf{x}_u + \mathbf{x}_v) \cdot (\mathbf{x}_u + \mathbf{x}_v) \\ &= (\mathbf{x}_u \cdot \mathbf{x}_u) du^2 + 2(\mathbf{x}_u \cdot \mathbf{x}_v) du dv + (\mathbf{x}_v \cdot \mathbf{x}_v) dv^2 \\ &= E du^2 + 2F du dv + G dv^2 \end{aligned} \tag{2.25}$$

where $E = \mathbf{x}_u \cdot \mathbf{x}_u$, $F = \mathbf{x}_u \cdot \mathbf{x}_v$, and $G = \mathbf{x}_v \cdot \mathbf{x}_v$.

Definition 2.5. The first fundamental form. The function $I(du, dv) = E du^2 + 2F du dv + G dv^2$ is called the first fundamental form of $\mathbf{x} = \mathbf{x}(u, v)$. E, F, and G are called the first fundamental coefficients, which are scalar-valued functions of u and v.

The first fundamental form has the following properties:

(1) I is a positive-definite quadratic form defined on (du, dv).
(2) I is independent of the representations though the coefficients E, F, and G are not.

For the curve $\mathbf{x}(t) = \mathbf{x}(u(t), v(t))$, $t_0 \leq t \leq t_1$, its length is the integral

$$\begin{aligned} s &= \int_{t_0}^{t_1} \left| \frac{d\mathbf{x}}{dt} \right| dt = \int_{t_0}^{t_1} \sqrt{\frac{d\mathbf{x}}{dt} \cdot \frac{d\mathbf{x}}{dt}} dt \\ &= \int_{t_0}^{t_1} \sqrt{E \left(\frac{du}{dt} \right)^2 + F \frac{du}{dt} \frac{dv}{dt} + G \left(\frac{dv}{dt} \right)^2} dt \end{aligned}$$

The angle α between u-curve and v-curve is

$$\alpha = \frac{\mathbf{x}_u \cdot \mathbf{x}_v}{|\mathbf{x}_u| |\mathbf{x}_v|} = \frac{F}{\sqrt{EG}}$$

It follows that the u-curve and v-curve are perpendicular at a point if and only if $F = 0$.

The area A of a region on $\mathbf{x} = \mathbf{x}(u, v)$, $(u, v) \in U$ is the double integral

$$A = \iint\limits_U \sqrt{EG - F^2}dudv$$

since $|\mathbf{x}_u \times \mathbf{x}_v|dudv = \sqrt{EG - F^2}dudv$ represents a first approximation of dA.

2.5.3 *The Second Fundamental Form*

The first fundamental form provides tools to measure curve lengths, angles, and areas on a surface, which are intrinsic properties of the surface. The second fundamental form is concerned with measurements of the 'flatness' of a surface at a point in the external 3D space.

As $\mathbf{x}_u$ and $\mathbf{x}_v$ are linearly independent, there exists a unit normal vector $\mathbf{N}$ such that

$$\mathbf{N} = \frac{\mathbf{x}_u \times \mathbf{x}_v}{|\mathbf{x}_u \times \mathbf{x}_v|}$$

Note $\mathbf{N}$ lies in the external 3D space where the surface is embedded. It follows that $d\mathbf{N} = \mathbf{N}_u du + \mathbf{N}_v dv$, which is orthogonal to $\mathbf{N}$ and lies in the tangent plane at $\mathbf{x}$ since $d\mathbf{N} \cdot \mathbf{N} = 0$, as shown in Fig. 2.12.

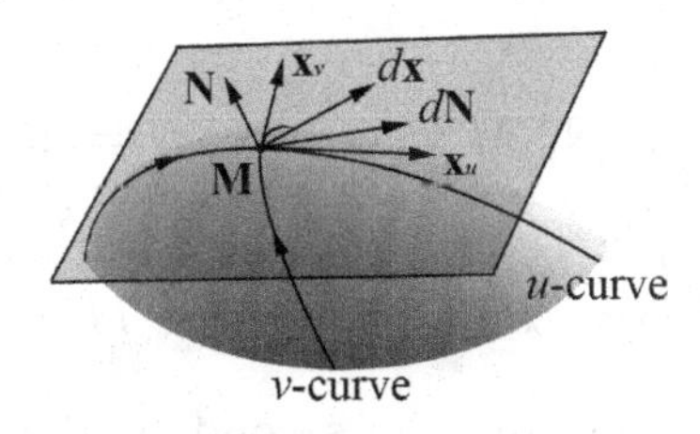

Fig. 2.12: d$\mathbf{N}$ and d$\mathbf{x}$ in the tangent plane at point M

We consider the quantity

$$
\begin{aligned}
II &= -d\mathbf{x} \cdot d\mathbf{N} = -(\mathbf{x}_u + \mathbf{x}_v) \cdot (\mathbf{N}_u + \mathbf{N}_v) \\
&= -(\mathbf{x}_u \cdot \mathbf{N}_u)du^2 - (\mathbf{x}_u \cdot \mathbf{N}_v + \mathbf{x}_v \cdot \mathbf{N}_u)dudv - (\mathbf{x}_v \cdot \mathbf{N}_v)dv^2 \\
&= Ldu^2 + 2Mdudv + Ndv^2 \qquad (2.26)
\end{aligned}
$$

where $L = -\mathbf{x}_u \cdot \mathbf{N}_u$, $M = -\frac{1}{2}(\mathbf{x}_u \cdot \mathbf{N}_v + \mathbf{x}_v \cdot \mathbf{N}_u)$, and $N = \mathbf{x}_v \cdot \mathbf{N}_v$.

Definition 2.6. The second fundamental form. The function $II(du, dv) = Ldu^2+2Mdudv+Ndv^2$ is called the second fundamental form of $\mathbf{x} = \mathbf{x}(u, v)$. The coefficients L, M, and N are called the second fundamental coefficients, which are scalar-valued functions of u and v.

The coefficients L, M, and N can also be obtained alternatively. Since both $\mathbf{x}_u$ and $\mathbf{x}_v$ are orthogonal to $\mathbf{N}$, it follows that

$$
\begin{aligned}
0 &= (\mathbf{x}_u \cdot \mathbf{N})_u = \mathbf{x}_{uu} \cdot \mathbf{N} + \mathbf{x}_u \cdot \mathbf{N}_u \\
0 &= (\mathbf{x}_u \cdot \mathbf{N})_v = \mathbf{x}_{uv} \cdot \mathbf{N} + \mathbf{x}_u \cdot \mathbf{N}_v \\
0 &= (\mathbf{x}_v \cdot \mathbf{N})_u = \mathbf{x}_{uv} \cdot \mathbf{N} + \mathbf{x}_v \cdot \mathbf{N}_u \\
0 &= (\mathbf{x}_v \cdot \mathbf{N})_v = \mathbf{x}_{vv} \cdot \mathbf{N} + \mathbf{x}_v \cdot \mathbf{N}_v
\end{aligned}
$$

Hence,

$$
\begin{aligned}
\mathbf{x}_{uu} \cdot \mathbf{N} &= -\mathbf{x}_u \cdot \mathbf{N}_u \\
\mathbf{x}_{uv} \cdot \mathbf{N} &= -\mathbf{x}_u \cdot \mathbf{N}_v = -\mathbf{x}_v \cdot \mathbf{N}_u \\
\mathbf{x}_{vv} \cdot \mathbf{N} &= -\mathbf{x}_v \cdot \mathbf{N}_v
\end{aligned}
$$

Alternative expressions for L, M, and N can be obtained respectively as

$$
L = \mathbf{x}_{uu} \cdot \mathbf{N},\ M = \mathbf{x}_{uv} \cdot \mathbf{N},\ N = \mathbf{x}_{vv} \cdot \mathbf{N} \qquad (2.27)
$$

Note that $d^2\mathbf{x} = \mathbf{x}_{uu}du^2 + 2\mathbf{x}_{uv}dudv + \mathbf{x}_{vv}dv^2$. It follows that the second fundamental form can be written as $II = d^2\mathbf{x} \cdot \mathbf{N}$.

Remark 2.7. Geometrically, the second fundamental form determines an osculating paraboloid in the neighborhood of a point, the shape of which could be elliptic, hyperbolic, parabolic, and planar depending on the values of L, M, and N.

2.5.4 *The Gauss–Weingarten Equations*

On an oriented regular surface $\mathbf{x} = \mathbf{x}(u, v)$, the vectors $\mathbf{x}_u$, $\mathbf{x}_v$, $\mathbf{N}$ are linearly independent at each point. We can thus express the derivatives $\mathbf{x}_{uu}$, $\mathbf{x}_{uv}$, $\mathbf{x}_{vv}$, $\mathbf{N}_u$, and $\mathbf{N}_v$ w.r.t. these three vectors as

$$\begin{aligned}
\mathbf{x}_{uu} &= \Gamma^1_{11}\mathbf{x}_u + \Gamma^2_{11}\mathbf{x}_v + \alpha_{11}\mathbf{N} \\
\mathbf{x}_{uv} &= \Gamma^1_{12}\mathbf{x}_u + \Gamma^2_{12}\mathbf{x}_v + \alpha_{12}\mathbf{N} \\
\mathbf{x}_{vv} &= \Gamma^1_{22}\mathbf{x}_u + \Gamma^2_{22}\mathbf{x}_v + \alpha_{22}\mathbf{N} \\
\mathbf{N}_u &= W^1_1\mathbf{x}_u + W^2_1\mathbf{x}_v \\
\mathbf{N}_v &= W^1_2\mathbf{x}_u + W^2_2\mathbf{x}_v
\end{aligned} \tag{2.28}$$

Note that $\mathbf{N}_u$ and $\mathbf{N}_v$ do not have components in the $\mathbf{N}$ direction as $\mathbf{N}$ is of unit length.

We now determine the coefficients Γ^i_{jk}, α_{ij}, and W^i_j in terms of the first fundamental coefficients — E, F, G — and the second fundamental coefficients —L, M, N.

Taking the dot product of both sides of the first three equations in Eq. (2.28) with $\mathbf{N}$ yields

$$\begin{aligned}
L &= \mathbf{x}_{uu} \cdot \mathbf{N} = \Gamma^1_{11}\mathbf{x}_u \cdot \mathbf{N} + \Gamma^2_{11}\mathbf{x}_v \cdot \mathbf{N} + \alpha_{11}\mathbf{N} \cdot \mathbf{N} \\
M &= \mathbf{x}_{uv} \cdot \mathbf{N} = \Gamma^1_{12}\mathbf{x}_u \cdot \mathbf{N} + \Gamma^2_{12}\mathbf{x}_v \cdot \mathbf{N} + \alpha_{12}\mathbf{N} \cdot \mathbf{N} \\
N &= \mathbf{x}_{vv} \cdot \mathbf{N} = \Gamma^1_{22}\mathbf{x}_u \cdot \mathbf{N} + \Gamma^2_{22}\mathbf{x}_v \cdot \mathbf{N} + \alpha_{22}\mathbf{N} \cdot \mathbf{N}
\end{aligned}$$

It follows that $\alpha_{11} = L$, $\alpha_{12} = M$, and $\alpha_{22} = N$ since $\mathbf{x}_u \cdot \mathbf{N} = 0$, $\mathbf{x}_v \cdot \mathbf{N} = 0$, and $\mathbf{N} \cdot \mathbf{N} = 1$. Taking the dot product of both sides of the first three equations in Eq. (2.28) with $\mathbf{x}_u$ and $\mathbf{x}_v$ yields the following system of six linear equations in six unknowns Γ^i_{jk}:

$$\begin{aligned}
\Gamma^1_{11}E + \Gamma^2_{11}F &= \mathbf{x}_{uu} \cdot \mathbf{x}_u = \frac{1}{2}E_u \\
\Gamma^1_{11}F + \Gamma^2_{11}G &= \mathbf{x}_{uu} \cdot \mathbf{x}_v = F_u - \frac{1}{2}E_v
\end{aligned}$$

$$\Gamma^1_{12}E + \Gamma^2_{12}F = \mathbf{x}_{uv} \cdot \mathbf{x}_u = \frac{1}{2}E_v$$

$$\Gamma^1_{12}F + \Gamma^2_{12}G = \mathbf{x}_{uv} \cdot \mathbf{x}_v = \frac{1}{2}G_u$$

$$\Gamma^1_{22}E + \Gamma^2_{22}F = \mathbf{x}_{vv} \cdot \mathbf{x}_u = F_v - \frac{1}{2}G_u$$

$$\Gamma^1_{22}F + \Gamma^2_{22}G = \mathbf{x}_{vv} \cdot \mathbf{x}_v = \frac{1}{2}G_v$$

Solving this system of linear equations for Γ^i_{jk} gives

$$\begin{aligned}
\Gamma^1_{11} &= \frac{GE_u - 2FF_u + FE_v}{2(EG - F^2)}, \quad \Gamma^2_{11} = \frac{2EF_u - EE_v - FE_u}{2(EG - F^2)} \\
\Gamma^1_{12} &= \frac{GE_v - FG_u}{2(EG - F^2)}, \quad \Gamma^2_{12} = \frac{EG_u - FE_v}{2(EG - F^2)} \\
\Gamma^1_{22} &= \frac{2GF_v - GG_u - FG_v}{2(EG - F^2)}, \quad \Gamma^2_{22} = \frac{EG_v - 2FF_v + FG_u}{2(EG - F^2)}
\end{aligned} \tag{2.29}$$

Taking the dot product of both sides of the last two equations in Eq. (2.28) with $\mathbf{x}_u$ and $\mathbf{x}_v$ yields the following system of four linear equations in four unknowns W^i_j:

$$W^1_1 E + W^2_1 F = \mathbf{N}_u \cdot \mathbf{x}_u = -L$$

$$W^1_1 F + W^2_1 G = \mathbf{N}_u \cdot \mathbf{x}_v = -M$$

$$W^1_2 E + W^2_2 F = \mathbf{N}_v \cdot \mathbf{x}_u = -M$$

$$W^1_2 F + W^2_2 G = \mathbf{N}_v \cdot \mathbf{x}_v = -N$$

Solving this system of linear equations for W^i_j gives

$$\begin{aligned}
W^1_1 &= \frac{MF - LG}{EG - F^2}, \quad W^2_1 = \frac{LF - ME}{EG - F^2} \\
W^1_2 &= \frac{NF - MG}{EG - F^2}, \quad W^2_2 = \frac{MF - NE}{EG - F^2}
\end{aligned} \tag{2.30}$$

Definition 2.7. The Gauss–Weingarten equations. The first three equations in Eq. (2.28) are called the Gauss equations, and the last two equations are called the Weingarten equations. The quantities Γ^i_{jk} are called the Christoffel symbols, which are symmetric relative to the lower indices — $\Gamma^1_{12} = \Gamma^1_{21}$, $\Gamma^2_{12} = \Gamma^1_{21}$ — since $\mathbf{x}_{uv} = \mathbf{x}_{vu}$. The quantities W^j_i are called the Weingarten coefficients.

2.6 The Moving-Frame Method Along the Coordinate Curves

Let a surface S be parameterized by $\mathbf{x} = \mathbf{x}(u, v)$ in such a way that its two coordinate curves — u-curve and v-curve — are perpendicular at any point. There exists the moving frame $(\mathbf{M}, \mathbf{e}_u, \mathbf{e}_v, \mathbf{e}_3)$, where $\mathbf{M}$ is the vector representing the point M on S, $\mathbf{e}_u$ is the unit vector along the u-curve, $\mathbf{e}_v$ is the unit vector along the v-curve, and $\mathbf{e}_3$ is the normal to the surface, as shown in Fig. 2.13.

It follows from Eqs. (2.1) and (2.2) that the infinitesimal displacement of the frame constitutes the following equations:

$$d\mathbf{M} = \omega_1 \mathbf{e}_u + \omega_2 \mathbf{e}_v$$
$$d\begin{bmatrix} \mathbf{e}_u \\ \mathbf{e}_v \\ \mathbf{e}_3 \end{bmatrix} = \begin{bmatrix} 0 & \omega_{12} & \omega_{13} \\ -\omega_{12} & 0 & \omega_{23} \\ -\omega_{12} & -\omega_{23} & 0 \end{bmatrix} \begin{bmatrix} \mathbf{e}_u \\ \mathbf{e}_v \\ \mathbf{e}_3 \end{bmatrix} \tag{2.31}$$

where ω_1, ω_2, ω_{12}, ω_{13}, and ω_{23} are the differential forms of the moving frame $(\mathbf{M}, \mathbf{e}_u, \mathbf{e}_v, \mathbf{e}_3)$. Note that $\omega_3 = 0$ since the component of $d\mathbf{M}$ in the direction of $\mathbf{e}_3$ is 0.

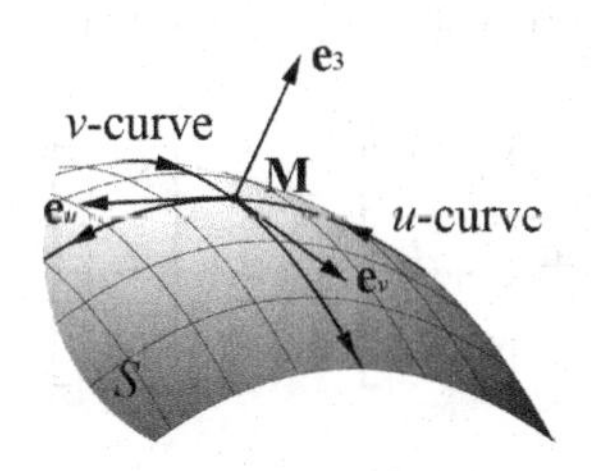

Fig. 2.13: The Darboux frame along the u-curve

Remark 2.8. These differential forms ω_1, ω_2, ω_{12}, ω_{13}, and ω_{23} represent the infinitesimal displacement of the frame $(\mathbf{M}, \mathbf{e}_u, \mathbf{e}_v, \mathbf{e}_3)$. Each of them is a scalar function of the variables u and v.

The relationships between $\mathbf{e}_u$ and $\mathbf{x}_u$ and $\mathbf{e}_v$ and $\mathbf{x}_v$ are, respectively,

$$\mathbf{e}_u = \frac{\mathbf{x}_u}{\sqrt{\mathbf{x}_u \cdot \mathbf{x}_u}} = \frac{\mathbf{x}_u}{\sqrt{E}}$$
$$\mathbf{e}_v = \frac{\mathbf{x}_v}{\sqrt{\mathbf{x}_v \cdot \mathbf{x}_v}} = \frac{\mathbf{x}_v}{\sqrt{G}}$$

where E and G are the first fundamental coefficients. It follows that

$$\omega_1 = \sqrt{E}du, \ \omega_2 = \sqrt{G}dv \tag{2.32}$$

Let $\mathbf{N}$ denote the normal at the point M, i.e., $\mathbf{N} = \mathbf{e}_3$. It follows that

$$\begin{bmatrix} \mathbf{x}_u \\ \mathbf{x}_v \\ \mathbf{N} \end{bmatrix} = \mathbf{A} \begin{bmatrix} \mathbf{e}_u \\ \mathbf{e}_v \\ \mathbf{e}_3 \end{bmatrix} \tag{2.33}$$

where

$$\mathbf{A} = \begin{bmatrix} \sqrt{E} & 0 & 0 \\ 0 & \sqrt{G} & 0 \\ 0 & 0 & 1 \end{bmatrix}$$

Differentiating the left side of Eq. (2.33) yields

$$d\begin{bmatrix} \mathbf{x}_u \\ \mathbf{x}_v \\ \mathbf{N} \end{bmatrix} = \begin{bmatrix} \mathbf{x}_{uu}du + \mathbf{x}_{uv}dv \\ \mathbf{x}_{uv}du + \mathbf{x}_{vv}dv \\ \mathbf{N}_u du + \mathbf{N}_v dv \end{bmatrix} = du \begin{bmatrix} \mathbf{x}_{uu} \\ \mathbf{x}_{uv} \\ \mathbf{N}_u \end{bmatrix} + dv \begin{bmatrix} \mathbf{x}_{uv} \\ \mathbf{x}_{vv} \\ \mathbf{N}_v \end{bmatrix} \tag{2.34}$$

The second-order derivatives of $\mathbf{x}$ — $\mathbf{x}_{uu}$, $\mathbf{x}_{uv}$, $\mathbf{x}_{vv}$ — and the first-order derivatives of $\mathbf{N}$ — $\mathbf{N}_u$, $\mathbf{N}_v$ — can be written in matrix form from the Gauss–Weingarten equations in Eqs. (2.29) and (2.30) as

$$\begin{aligned}
\begin{bmatrix} \mathbf{x}_{uu} \\ \mathbf{x}_{uv} \\ \mathbf{N}_u \end{bmatrix} &= \begin{bmatrix} \Gamma_{11}^1 & \Gamma_{11}^2 & L \\ \Gamma_{12}^1 & \Gamma_{12}^2 & M \\ W_1^1 & W_1^2 & 0 \end{bmatrix} \begin{bmatrix} \mathbf{x}_u \\ \mathbf{x}_v \\ \mathbf{N} \end{bmatrix} \\
\begin{bmatrix} \mathbf{x}_{uv} \\ \mathbf{x}_{vv} \\ \mathbf{N}_v \end{bmatrix} &= \begin{bmatrix} \Gamma_{12}^1 & \Gamma_{12}^2 & M \\ \Gamma_{22}^1 & \Gamma_{22}^2 & N \\ W_2^1 & W_2^2 & 0 \end{bmatrix} \begin{bmatrix} \mathbf{x}_u \\ \mathbf{x}_v \\ \mathbf{N} \end{bmatrix}
\end{aligned} \tag{2.35}$$

Substituting Eq. (2.35) into Eq. (2.34) yields

$$d\begin{bmatrix} \mathbf{x}_u \\ \mathbf{x}_v \\ \mathbf{N} \end{bmatrix} = \left(du \begin{bmatrix} \Gamma_{11}^1 & \Gamma_{11}^2 & L \\ \Gamma_{12}^1 & \Gamma_{12}^2 & M \\ W_1^1 & W_1^2 & 0 \end{bmatrix} + dv \begin{bmatrix} \Gamma_{12}^1 & \Gamma_{12}^2 & M \\ \Gamma_{22}^1 & \Gamma_{22}^2 & N \\ W_2^1 & W_2^2 & 0 \end{bmatrix} \right) \mathbf{A} \begin{bmatrix} \mathbf{e}_u \\ \mathbf{e}_v \\ \mathbf{e}_3 \end{bmatrix} \tag{2.36}$$

Differentiating the right side of Eq. (2.33) yields

$$d\left(\mathbf{A} \begin{bmatrix} \mathbf{e}_u \\ \mathbf{e}_v \\ \mathbf{e}_3 \end{bmatrix} \right) = \left(d\mathbf{A} + \mathbf{A} \begin{bmatrix} 0 & \omega_{12} & \omega_{13} \\ -\omega_{12} & 0 & \omega_{23} \\ -\omega_{13} & -\omega_{23} & 0 \end{bmatrix} \right) \begin{bmatrix} \mathbf{e}_u \\ \mathbf{e}_v \\ \mathbf{e}_3 \end{bmatrix} \tag{2.37}$$

where

$$d\mathbf{A} = \begin{bmatrix} \frac{E_u du + E_v dv}{2\sqrt{E}} & 0 & 0 \\ 0 & \frac{G_u du + G_v dv}{2\sqrt{G}} & 0 \\ 0 & 0 & 0 \end{bmatrix}$$

With the Christoffel symbols and Weingarten coefficients in terms of the first and second fundamental coefficients in Eqs. (2.29) and (2.30), equating Eqs. (2.36) and (2.37) yields six scalar equations, from which the differential forms ω_{12}, ω_{13}, and ω_{23} can be obtained as

$$\begin{aligned}
\omega_{12} &= \frac{-E_v du + G_u dv}{2\sqrt{EG}} \\
\omega_{13} &= \frac{L du + M dv}{\sqrt{E}} \\
\omega_{23} &= \frac{M du + N dv}{\sqrt{G}}
\end{aligned} \tag{2.38}$$

2.7 The Moving-Frame Method Along an Arbitrary Curve

Suppose there exists another moving frame ($\mathbf{M}$, $\mathbf{e}_1$, $\mathbf{e}_2$, $\mathbf{e}_3$) that can be obtained by rotating the frame ($\mathbf{M}, \mathbf{e}_u, \mathbf{e}_v, \mathbf{e}_3$) by an angle φ about $\mathbf{e}_3$, as shown in Fig. 2.14.

The relationship between the two frames can be expressed as

$$\begin{bmatrix} \mathbf{e}_1 \\ \mathbf{e}_2 \\ \mathbf{e}_3 \end{bmatrix} = \mathbf{R} \begin{bmatrix} \mathbf{e}_u \\ \mathbf{e}_v \\ \mathbf{e}_3 \end{bmatrix} \tag{2.39}$$

where

$$\mathbf{R} = \begin{bmatrix} \cos\varphi & \sin\varphi & 0 \\ -\sin\varphi & \cos\varphi & 0 \\ 0 & 0 & 1 \end{bmatrix}$$

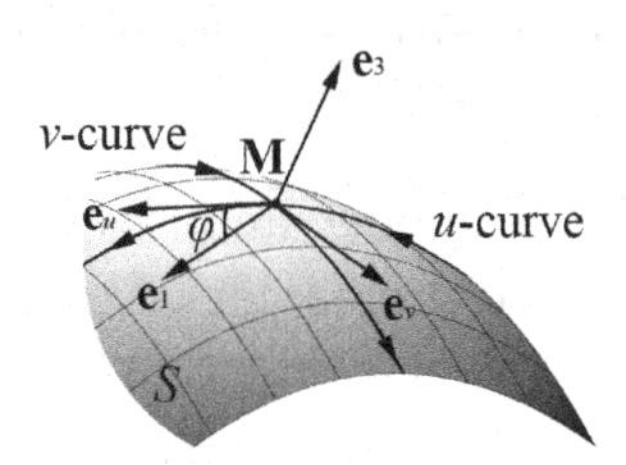

Fig. 2.14: The relationship between the moving frames ($\mathbf{M}$, $\mathbf{e}_1$, $\mathbf{e}_2$, $\mathbf{e}_3$) and ($\mathbf{M}$, $\mathbf{e}_u$, $\mathbf{e}_v$, $\mathbf{e}_3$)

Differentiating the left side of Eq. (2.39) yields

$$d\left(\begin{bmatrix}\mathbf{e}_1\\\mathbf{e}_2\\\mathbf{e}_3\end{bmatrix}\right)=\begin{bmatrix}0&\omega'_{12}&\omega'_{13}\\-\omega'_{12}&0&\omega'_{23}\\-\omega'_{13}&-\omega'_{23}&0\end{bmatrix}\begin{bmatrix}\mathbf{e}_1\\\mathbf{e}_2\\\mathbf{e}_3\end{bmatrix}$$
$$=\begin{bmatrix}0&\omega'_{12}&\omega'_{13}\\-\omega'_{12}&0&\omega'_{23}\\-\omega'_{13}&-\omega'_{23}&0\end{bmatrix}\mathbf{R}\begin{bmatrix}\mathbf{e}_u\\\mathbf{e}_v\\\mathbf{e}_3\end{bmatrix} \tag{2.40}$$

where ω'_{12}, ω'_{13}, and ω'_{23} are the differential forms of the moving frame $(\mathbf{M},\ \mathbf{e}_1,\ \mathbf{e}_2,\ \mathbf{e}_3)$.

Differentiating the right side of Eq. (2.39) yields

$$d\left(\mathbf{R}\begin{bmatrix}\mathbf{e}_u\\\mathbf{e}_v\\\mathbf{e}_3\end{bmatrix}\right)=\left(d\mathbf{R}+\mathbf{R}\begin{bmatrix}0&\omega_{12}&\omega_{13}\\-\omega_{12}&0&\omega_{23}\\-\omega_{13}&-\omega_{23}&0\end{bmatrix}\right)\begin{bmatrix}\mathbf{e}_u\\\mathbf{e}_v\\\mathbf{e}_3\end{bmatrix} \tag{2.41}$$

where ω_{12}, ω_{13}, and ω_{23} are the differential forms of the frame $(\mathbf{M},\mathbf{e}_u,\mathbf{e}_v,\mathbf{e}_3)$ given in Eq. (2.31).

Equating Eqs. (2.40) and (2.41) yields eight scalar equations, from which the differential forms ω'_{12}, ω'_{13}, and ω'_{23} can be obtained as

$$\begin{aligned}\omega'_{12}&=\omega_{12}+d\varphi\\\omega'_{13}&=\omega_{13}\cos\varphi+\omega_{23}\sin\varphi\\\omega'_{23}&=-\omega_{13}\sin\varphi+\omega_{23}\cos\varphi\end{aligned} \tag{2.42}$$

In particular, the moving frame $(\mathbf{M},\mathbf{e}_v,-\mathbf{e}_u,\mathbf{e}_3)$ can be obtained by rotating the frame $(\mathbf{M},\ \mathbf{e}_u,\mathbf{e}_v,\mathbf{e}_3)$ about $\mathbf{e}_3$ by an angle of $\pi/2$, as shown in Fig. 2.15.

The differential forms ω'_{12}, ω'_{13}, and ω'_{23} of the moving frame $(\mathbf{M},\mathbf{e}_v,-\mathbf{e}_u,\mathbf{e}_3)$ can be obtained from Eqs. (2.38) and (2.42) by

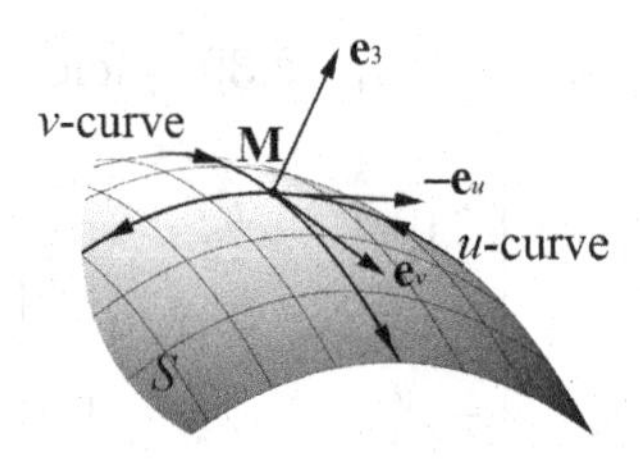

Fig. 2.15: The moving frame $(\mathbf{M}, \mathbf{e}_v, -\mathbf{e}_u, \mathbf{e}_3)$ along the v-curve

setting $\varphi = \pi/2$ as

$$\begin{aligned} \omega'_{12} &= \frac{-E_v du + G_u dv}{2\sqrt{EG}} \\ \omega'_{13} &= \frac{M du + N dv}{\sqrt{G}} \\ \omega'_{23} &= -\frac{L du + M dv}{\sqrt{E}} \end{aligned} \tag{2.43}$$

2.8 The Darboux Frame and the Darboux Formulas of an Oriented Curve on a Surface

Let $\mathbf{x}(t) = \mathbf{x}(u(t), v(t))$ represent a curve L traced on the surface S. We attach a moving frame $(\mathbf{M}, \mathbf{e}_1, \mathbf{e}_2, \mathbf{e}_3)$ to a point M of L in such a way that $\mathbf{e}_1$ tangent to L, $\mathbf{e}_3$ normal to the surface, and $\mathbf{e}_2 = \mathbf{e}_3 \times \mathbf{e}_1$, as shown in Fig. 2.16.

Definition 2.8. The Darboux frame. The moving frame $(\mathbf{M}, \mathbf{e}_1, \mathbf{e}_2, \mathbf{e}_3)$ defined above is called the Darboux frame along the oriented curve L on the surface S.

The equations of the infinitesimal displacement of the Darboux frame are

$$\begin{aligned} d\mathbf{M} &= \omega_1 \mathbf{e}_1 \\ d\begin{bmatrix} \mathbf{e}_1 \\ \mathbf{e}_2 \\ \mathbf{e}_3 \end{bmatrix} &= \begin{bmatrix} 0 & \omega_{12} & \omega_{13} \\ -\omega_{12} & 0 & \omega_{23} \\ \omega_{13} & -\omega_{23} & 0 \end{bmatrix} \begin{bmatrix} \mathbf{e}_1 \\ \mathbf{e}_2 \\ \mathbf{e}_3 \end{bmatrix} \end{aligned} \tag{2.44}$$

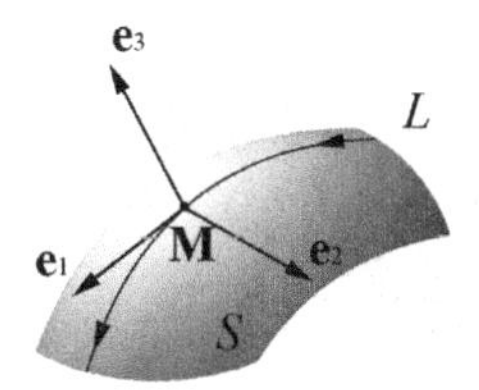

Fig. 2.16: The Darboux frame ($\mathbf{M}$, $\mathbf{e}_1$, $\mathbf{e}_2$, $\mathbf{e}_3$) on the point M of the curve L

where ω_1, ω_{12}, ω_{12}, and ω_{12} are differential forms of the parameter t. Similar to the Frenet formulas, we have $\omega_1 = ds$, where s represents arch length of the curve L.

We introduce three functions on the curve L: the geodesic curvature k_g, the normal curvature k_n, and the geodesic torsion τ_g, which are defined respectively as

$$k_g = \frac{\omega_{12}}{\omega_1}, \quad k_n = \frac{\omega_{13}}{\omega_1}, \quad \tau_g = \frac{\omega_{23}}{\omega_1} \tag{2.45}$$

Definition 2.9. The Darboux formulas. With the introduction of arc length s, the geodesic curvature k_g, the normal curvature k_n, and the geodesic torsion τ_g, the Darboux formulas are defined as

$$\begin{aligned} \frac{d\mathbf{M}}{ds} &= \mathbf{e}_1 \\ \frac{d}{ds}\begin{bmatrix} \mathbf{e}_1 \\ \mathbf{e}_2 \\ \mathbf{e}_3 \end{bmatrix} &= \begin{bmatrix} 0 & k_g & k_n \\ -k_g & 0 & \tau_g \\ -k_n & -\tau_g & 0 \end{bmatrix} \begin{bmatrix} \mathbf{e}_1 \\ \mathbf{e}_2 \\ \mathbf{e}_3 \end{bmatrix} \end{aligned} \tag{2.46}$$

2.9 Geometry of the Geodesic Curvature, Normal Curvature, and Geodesic Torsion

We examine the geometry of the geodesic k_g, normal curvature k_n, and geodesic torsion τ_g in relation to the curvature k and the torsion τ of the curve L.

Let $\mathbf{e}_{iF}$ denote the base vectors of Frenet frame and $\mathbf{e}_{iD}$ the base vectors of the Darboux frame, then $\mathbf{e}_{1F} = \mathbf{e}_{1D}$, as shown in Fig. 2.17.

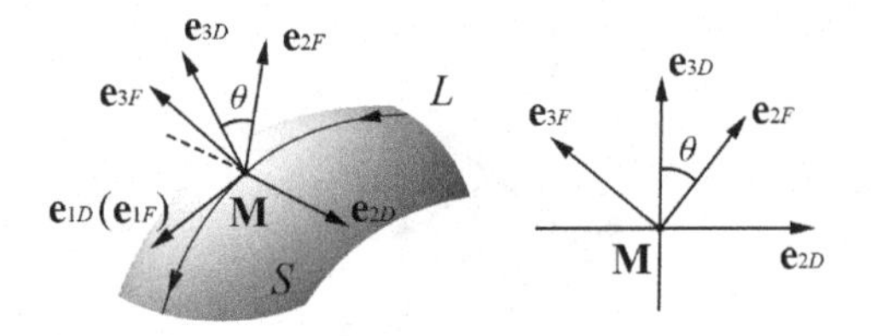

Fig. 2.17: The Darboux frame and the Frenet frame at the point M of the curve L

It follows that $d\mathbf{e}_{1F}/ds = d\mathbf{e}_{1D}/ds$, which gives

$$\frac{d\mathbf{e}_{1D}}{ds} = k_g\mathbf{e}_{2D} + k_n\mathbf{e}_{3D} = k\mathbf{e}_{2F} = \frac{d\mathbf{e}_{1F}}{ds} \tag{2.47}$$

Taking the dot product of both sides of Eq. (2.47) with $\mathbf{e}_{3D}$ yields

$$k_n = k(\mathbf{e}_{2F} \cdot \mathbf{e}_{3D}) = k\cos\theta \tag{2.48}$$

where θ is the angle between $\mathbf{e}_{3D}$ and $\mathbf{e}_{2F}$. Similarly, the geodesic curvature k_g can be obtained as

$$k_g = k(\mathbf{e}_{2F} \cdot \mathbf{e}_{2D}) = k\sin\theta \tag{2.49}$$

It is clear that

$$\mathbf{e}_{2F} = \sin\theta\mathbf{e}_{2D} + \cos\theta\mathbf{e}_{3D} \tag{2.50}$$

Taking the derivative of the left side of Eq. (2.50) w.r.t. s yields

$$\frac{d\mathbf{e}_{2F}}{ds} = -k\mathbf{e}_{1F} + \tau\mathbf{e}_{3F} \tag{2.51}$$

Taking the derivative of the right side of Eq. (2.50) w.r.t. s yields

$$\begin{aligned}\frac{d}{ds}(\sin\theta\mathbf{e}_{2D} + \cos\theta\mathbf{e}_{3D}) &= \cos\theta\frac{d\theta}{ds}\mathbf{e}_{2D} + \sin\theta(-k_g\mathbf{e}_{1D} + \tau_g\mathbf{e}_{3D})\\ &\quad - \sin\theta\frac{d\theta}{ds}\mathbf{e}_{3D} + \cos\theta(-k_n\mathbf{e}_{1D} - \tau_g\mathbf{e}_{2D})\end{aligned} \tag{2.52}$$

Substituting $\mathbf{e}_{1F}$ with $\mathbf{e}_{1D}$ and $\mathbf{e}_{3F}$ with $(-\cos\theta\mathbf{e}_{2D} + \sin\theta\mathbf{e}_{3D})$ and equating Eqs. (2.51) and (2.52) yield the following three scalar equations:

$$\begin{aligned} &k_n \cos\theta + k_g \sin\theta - k = 0 \\ &-\cos\theta \left(\frac{d\theta}{ds} + \tau - \tau_g \right) = 0 \\ &\sin\theta \left(\frac{d\theta}{ds} + \tau - \tau_g \right) = 0 \end{aligned} \tag{2.53}$$

The first equation manifests itself. The second and third equations give

$$\tau_g = \tau + \frac{d\theta}{ds} \tag{2.54}$$

Remark 2.9. The properties of a surface can be partially revealed through studying the curves on it. Some curves have special continuous geometric characteristics. For instance, a line of curvature is a regular connected curve such that the geodesic torsion τ_g vanishes and the normal curvature k_n takes either maximum or minimum value along this curve. A geodesic is a regular connected curve such that the geodesic curvature k_g vanishes and is a generalized notion of straight line (shortest path) in a curved space.

Example 2.6. A planar curve L with curvature k lies in a plane S, as shown in Fig. 2.18. Obtain the geodesic curvature k_g, normal curvature k_n, and geodesic torsion τ_g of this curve.

Solution:
The plane S may be parameterized as

$$\mathbf{x}(u, v) = \begin{bmatrix} u \; v \; 0 \end{bmatrix}$$

Let $\mathbf{e}_3$ represent the normal to the surface S. We choose it to be pointing upwards, i.e., $\mathbf{e}_3 = \mathbf{x}_u \times \mathbf{x}_v / |\mathbf{x}_u \times \mathbf{x}_v| = \mathbf{k}$.

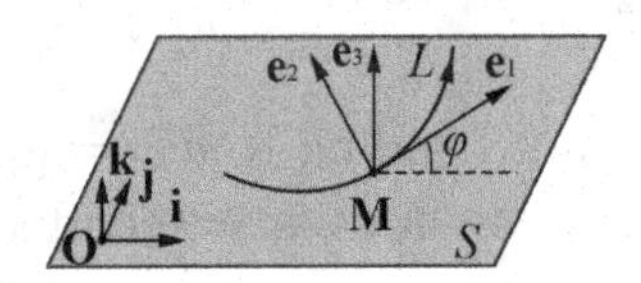

Fig. 2.18: The Darboux frame of the planar curve L

Let M denote a point on the curve L, which can be parameterized as a function of time t as

$$\mathbf{M}(t) = \mathbf{x}(u(t), v(t)) = \left[u(t)\ v(t)\ 0\right]$$

where both u and v are functions of time t. Set up the Darboux frame $(\mathbf{M}, \mathbf{e}_1, \mathbf{e}_2, \mathbf{e}_3)$, where $\mathbf{e}_1$ is the unit tangent vector and $\mathbf{e}_2 = \mathbf{e}_3 \times \mathbf{e}_1$.

Differentiating $\mathbf{M}$ yields

$$d\mathbf{M} = \left[\dot{u}dt,\ \dot{v}dt,\ 0\right] = \omega_1 \mathbf{e}_1$$

where $\dot{u}$ represents du/dt and $\dot{v}$ represents dv/dt. It follows that

$$\omega_1 = |d\mathbf{M}| = \sqrt{\dot{u}^2 + \dot{v}^2}dt$$

$$\mathbf{e}_1 = \frac{\left[\dot{u},\ \dot{v},\ 0\right]}{\sqrt{\dot{u}^2 + \dot{v}^2}}$$

The vector $\mathbf{e}_2$ can be obtained as

$$\mathbf{e}_2 = \mathbf{e}_3 \times \mathbf{e}_1 = \frac{\left[-\dot{v},\ \dot{u},\ 0\right]}{\sqrt{\dot{u}^2 + \dot{v}^2}}$$

Differentiating the tangent vector $\mathbf{e}_1$ yields

$$d\mathbf{e}_1 = \left[-\frac{\dot{v}(\dot{u}\ddot{v} - \ddot{u}\dot{v})dt}{\sqrt{(\dot{u}^2 + \dot{v}^2)^3}},\ \frac{\dot{u}(\dot{u}\ddot{v} - \ddot{u}\dot{v})dt}{\sqrt{(\dot{u}^2 + \dot{v}^2)^3}},\ 0\right]$$

where $\ddot{u} = d^2u/dt^2$ and $\ddot{v} = d^2v/dt^2$. The differential form ω_{12} can be obtained as

$$\omega_{12} = d\mathbf{e}_1 \cdot \mathbf{e}_2 = \frac{(\dot{u}\ddot{v} - \ddot{u}\dot{v})dt}{\dot{u}^2 + \dot{v}^2}$$

The vector $\mathbf{e}_3$ is a constant, leading to $d\mathbf{e}_3 = [0, 0, 0]$. It follows that the differential forms $\omega_{13} = -d\mathbf{e}_3 \cdot \mathbf{e}_1 = 0$ and $\omega_{23} = -d\mathbf{e}_3 \cdot \mathbf{e}_2 = 0$.

Hence, the geodesic curvature k_g, normal curvature k_n, and geodesic torsion τ_g of the planar curve L can be obtained from

Eq. (2.45) respectively:

$$k_g = \frac{\omega_{12}}{\omega_1} = \frac{\dot{u}\ddot{v} - \ddot{u}\dot{v}}{\sqrt{(\dot{u}^2 + \dot{v}^2)^3}}$$

$$k_n = \frac{\omega_{13}}{\omega_1} = 0$$

$$\tau_g = \frac{\omega_{23}}{\omega_1} = 0$$

The Darboux formulas can be obtained straightforwardly from Eq. (2.46). □

Remark 2.10. It can be straightforwardly verified that the absolute value of k_g of the Darboux frame of a planar curve L is equal to the absolute value of the curvature k of the Frenet frame, i.e., $|k_g| = |k|$.

Example 2.7. Obtain the geodesic curvature k_g, normal curvature k_n, and geodesic torsion τ_g of a small circle on a unit sphere S, as shown in Fig. 2.19.

Solution:
The unit sphere S may be parameterized as

$$\mathbf{x}(u, v) = \begin{bmatrix}\cos u \cos v & -\cos u \sin v & \sin u\end{bmatrix}$$
$$u \in (-\pi/2, \pi/2), \ v \in [0, 2\pi)$$

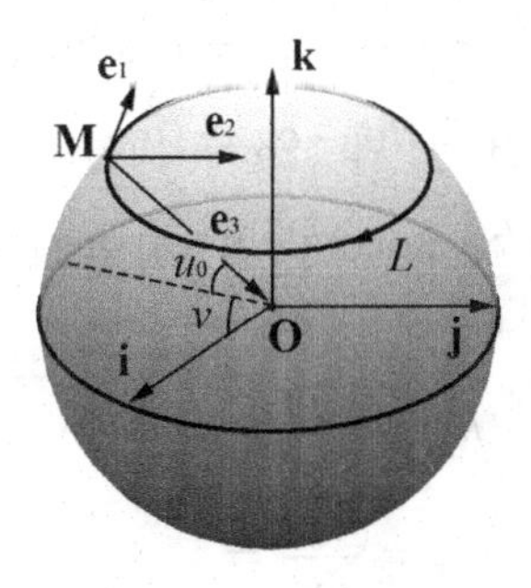

Fig. 2.19: The Darboux frame of a small circle L on a unit sphere S

The normal vector $\mathbf{e}_3$ of the sphere S is chosen to be pointing to the sphere center. It follows that

$$\mathbf{e}_3 = -\mathbf{x}_u \times \mathbf{x}_v / |\mathbf{x}_u \times \mathbf{x}_v| = \begin{bmatrix} -\cos u \cos v & \cos u \sin v & -\sin u \end{bmatrix}$$

Let M denote a point on the small circle L, which can be parameterized as a function of time t as

$$\mathbf{M}(t) = \begin{bmatrix} \cos u_0 \cos v(t) & -\cos u_0 \sin v(t) & \sin u_0 \end{bmatrix}$$

where u_0 is a constant and v is a function of time t. Set up the Darboux frame ($\mathbf{M}$, $\mathbf{e}_1$, $\mathbf{e}_2$, $\mathbf{e}_3$), where $\mathbf{e}_1$ is the unit tangent vector and $\mathbf{e}_2 = \mathbf{e}_3 \times \mathbf{e}_1$, as shown in Fig. 2.19.

The differentiation of $\mathbf{M}$ can be obtained as

$$d\mathbf{M} = \begin{bmatrix} -\cos u_0 \sin v dv, & -\cos u_0 \cos v dv & 0 \end{bmatrix} = \omega_1 \mathbf{e}_1$$

It follows that

$$\omega_1 = \cos u_0 dv$$

$$\mathbf{e}_1 = \begin{bmatrix} -\sin v & -\cos v & 0 \end{bmatrix}$$

Let s represent arc length of L, then $ds = \omega_1$. The vector $\mathbf{e}_2$ can be obtained as

$$\mathbf{e}_2 = \mathbf{e}_3 \times \mathbf{e}_1 = \begin{bmatrix} -\sin u_0 \cos v & \sin u_0 \sin v & \cos u_0 \end{bmatrix}$$

The differential forms ω_{12}, ω_{13}, and ω_{23} can be obtained respectively as

$$\omega_{12} = d\mathbf{e}_1 \cdot \mathbf{e}_2 = \sin u_0 dv$$

$$\omega_{13} = d\mathbf{e}_1 \cdot \mathbf{e}_3 = \cos u_0 dv$$

$$\omega_{23} = d\mathbf{e}_2 \cdot \mathbf{e}_3 = 0$$

The geodesic curvature k_g, normal curvature k_n, and geodesic torsion τ_g of L can be obtained respectively as

$$k_g = \frac{\omega_{12}}{\omega_1} = \tan u_0$$

$$k_n = \frac{\omega_{13}}{\omega_1} = 1$$

$$\tau_g = \frac{\omega_{23}}{\omega_1} = 0$$

The Darboux formulas can be obtained straightforwardly from Eq. (2.46). □

2.10 The Curvatures of a Surface Curve in Terms of the Curvatures of the Coordinate Curves

We derive the normal curvature k_n, geodesic curvature k_g, and geodesic torsion τ_g of a curve L traced on a surface $\mathbf{x} = \mathbf{x}(u, v)$ in terms of the curvatures of the coordinate curves.

The u-curve of a surface $\mathbf{x} = \mathbf{x}(u, v)$ can be parameterized as $\mathbf{x}(t) = \mathbf{x}(u(t), v)$, where $v = v_0$ is a constant. Let M denote a point on the u-curve; we set up the Darboux frame $(\mathbf{M}, \mathbf{e}_u, \mathbf{e}_v, \mathbf{e}_3)$ along this curve, as shown in Fig. 2.14.

Note that $dv = 0$ along the u-curve since v is a constant. The differential forms of the Darboux frame $(\mathbf{M}, \mathbf{e}_u, \mathbf{e}_v, \mathbf{e}_3)$ can be obtained from Eqs. (2.32) and (2.38) by setting $dv = 0$ as

$$\begin{aligned} \omega_1 &= \sqrt{E}du \\ \omega_{12} &= -\frac{E_v du}{2\sqrt{EG}} \\ \omega_{13} &= \frac{Ldu}{\sqrt{E}} \\ \omega_{23} &= \frac{Mdu}{\sqrt{G}} \end{aligned} \tag{2.55}$$

Let k_{gu} denote the geodesic curvature, k_{nu} the normal curvature, and τ_{gu} the geodesic torsion, of the u-curve. They can be obtained from Eq. (2.45) as

$$\begin{aligned} k_{gu} &= \frac{\omega_{12}}{\omega_1} = -\frac{E_v}{2E\sqrt{G}} \\ k_{nu} &= \frac{\omega_{13}}{\omega_1} = \frac{L}{E} \\ \tau_{gu} &= \frac{\omega_{23}}{\omega_1} = \frac{M}{\sqrt{EG}} \end{aligned} \tag{2.56}$$

Similarly, the v-curve on the surface S can be parameterized as $\mathbf{x}(t) = \mathbf{x}(u, v(t))$, where $u = u_0$ is a constant. Let M denote a point on the v-curve; we set up the Darboux frame $(\mathbf{M}, \mathbf{e}_v, -\mathbf{e}_u, \mathbf{e}_3)$ along this curve, as shown in Fig. 2.15.

Note that $du = 0$ along the v-curve since u is a constant. The differential forms of the Darboux frame $(\mathbf{M}, \mathbf{e}_v, -\mathbf{e}_u, \mathbf{e}_3)$ can be obtained from Eqs. (2.32) and (2.43) by setting $du = 0$ as

$$\begin{aligned} \omega_1' &= \sqrt{G}dv \\ \omega_{12}' &= \frac{G_u dv}{2\sqrt{EG}} \\ \omega_{13}' &= \frac{N dv}{\sqrt{G}} \\ \omega_{23}' &= -\frac{M dv}{\sqrt{E}} \end{aligned} \tag{2.57}$$

Let k_{gv}, k_{nv}, and τ_{gv} denote the geodesic curvature, normal curvature, and geodesic torsion of the v-curve, respectively. These entities can be obtained from Eq. (2.45) as

$$\begin{aligned} k_{gv} &= \frac{\omega_{12}'}{\omega_1'} = \frac{G_u}{2G\sqrt{E}} \\ k_{nv} &= \frac{\omega_{13}'}{\omega_1'} = \frac{N}{G} \\ \tau_{gv} &= \frac{\omega_{23}'}{\omega_1'} = -\frac{M}{\sqrt{EG}} \end{aligned} \tag{2.58}$$

A curve L traced on the surface S can be parameterized as $\mathbf{x}(t) = \mathbf{x}(u(t), v(t))$. Let φ denote the angle between L and the u-curve at the point M, and we set up the Darboux frame $(\mathbf{M}, \mathbf{e}_1, \mathbf{e}_2, \mathbf{e}_3)$ along the curve L, as shown in Fig. 2.20.

It follows that

$$d\mathbf{M} = \omega_{1L}\mathbf{e}_1 = \sqrt{E(du)^2 + G(dv)^2}\mathbf{e}_1 = \sqrt{\omega_1^2 + \omega_2^2}\mathbf{e}_1 \tag{2.59}$$

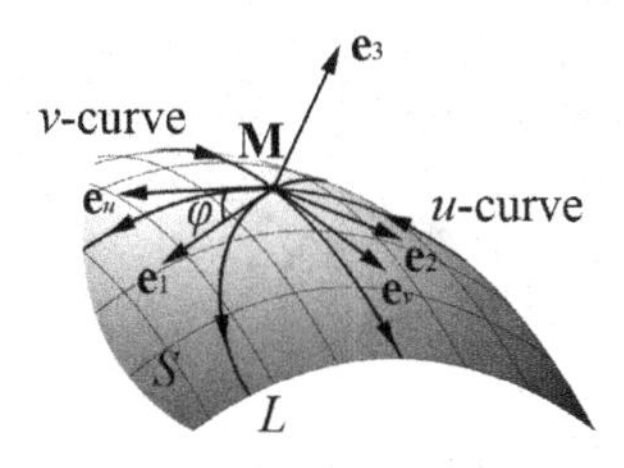

Fig. 2.20: The Darboux frame of the curve L on the surface S

where $\omega_1 = \sqrt{E}du$ and $\omega_2 = \sqrt{G}dv$, as in Eq. (2.32). It follows that

$$\omega_{1L} = \sqrt{\omega_1^2 + \omega_2^2} = ds \tag{2.60}$$

where s is the arc length of L. Since the angle between $\mathbf{e}_u$ and $\mathbf{e}_1$ is φ, it follows that

$$\begin{aligned} \cos\varphi &= \frac{\omega_1}{\omega_{1L}} = \frac{\omega_1}{ds} \\ \sin\varphi &= \frac{\omega_2}{\omega_{1L}} = \frac{\omega_2}{ds} \end{aligned} \tag{2.61}$$

The frame $(\mathbf{M}, \mathbf{e}_1, \mathbf{e}_2, \mathbf{e}_3)$ can be obtained by rotating the frame $(\mathbf{M}, \mathbf{e}_u, \mathbf{e}_v, \mathbf{e}_3)$ by the angle φ about $\mathbf{e}_3$. The differential forms ω_{12L}, ω'_{13L}, and ω_{23L} of the Darboux frame $(\mathbf{M}, \mathbf{e}_1, \mathbf{e}_2, \mathbf{e}_3)$ are given from Eq. (2.42).

Let k_g, k_n, and τ_g denote the geodesic curvature, normal curvature, and geodesic torsion of L, respectively. These geometric invariants can be obtained from Eqs. (2.32), (2.57), (2.58), and (2.61) after some algebraic manipulation as

$$\begin{aligned} k_g &= \frac{\omega_{12L}}{\omega_{1L}} = k_{gu}\cos\varphi + k_{gv}\sin\varphi + \frac{d\varphi}{ds} \\ k_n &= \frac{\omega_{13L}}{\omega_{1L}} = k_{nu}\cos^2\varphi + 2\tau_{gu}\cos\varphi\sin\varphi + k_{nv}\sin^2\varphi \\ \tau_g &= \frac{\omega_{23L}}{\omega_{1L}} = \tau_{gu}\cos 2\varphi + \frac{1}{2}(k_{nv} - k_{nu})\sin 2\varphi \end{aligned} \tag{2.62}$$

Remark 2.11. In Eq. (2.62), the expression of the geodesic curvature k_g is in accordance with the Liouville theorem by

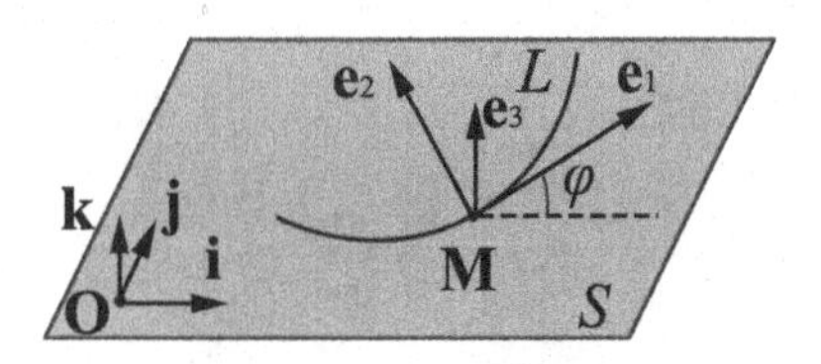

Fig. 2.21: The curvature in a known direction

Carmo (1976); the expression of the normal curvature k_n is a generalized Euler's equation by Cartan (1996).

Example 2.8. A planar curve L lies in a plane S, on which a frame ($\mathbf{O}$, $\mathbf{i}$, $\mathbf{j}$, $\mathbf{k}$) is fixed. Let s denote the arc length and k the curvature of L, and φ the angle between the unit tangent $\mathbf{e}_1$ and $\mathbf{i}$, as shown in Fig. 2.21. Show $k = d\varphi/ds$.

Solution:
The plane S can be parameterized as

$$\mathbf{x}(u, v) = \begin{bmatrix} u, \ v, \ 0 \end{bmatrix}$$

The coefficients of the first fundamental forms are, respectively

$$E = \mathbf{x}_u \cdot \mathbf{x}_u = 1, \ F = \mathbf{x}_u \cdot \mathbf{x}_v = 0, \ G = \mathbf{x}_v \cdot \mathbf{x}_v = 1$$

The coefficients of the second fundamental forms are, respectively

$$L = \mathbf{N} \cdot \mathbf{x}_{uu} = 0, \ M = \mathbf{N} \cdot \mathbf{x}_{uv} = 0, \ G = \mathbf{N} \cdot \mathbf{x}_{vv} = 0$$

It follows from Eqs. (2.56) and (2.58) that the curvatures of the u-curve and v-curve are, respectively

$$k_{gu} = k_{nu} = \tau_{gu} = 0, \ k_{gv} = k_{nv} = \tau_{gv} = 0$$

The curvatures of the curve L can be obtained from Eq. (2.62) as

$$k_g = \frac{d\varphi}{ds}, \ k_n = 0, \ \tau_g = 0$$

We have known from Example 2.5 that the geodesic curvature of a planar curve is its curvature. It follows that $k = k_g = d\varphi/ds$.

This can also be verified by the following steps. Recall that the Frenet formulas of a planar curve are

$$\frac{d}{ds}\begin{bmatrix}\mathbf{e}_1\\ \mathbf{e}_2\end{bmatrix} = \begin{bmatrix}0, & k\\ -k, & 0\end{bmatrix}\begin{bmatrix}\mathbf{e}_1\\ \mathbf{e}_2\end{bmatrix}$$

The unit tangent $\mathbf{e}_1$ making an angle φ with $\mathbf{i}$ implies

$$\cos\varphi = \mathbf{e}_1 \cdot \mathbf{i}$$

Differentiating both sides of the above equation w.r.t. the arc length s yields

$$-\sin\varphi\frac{d\varphi}{ds} = \frac{d\mathbf{e}_1}{ds}\cdot\mathbf{i} = k\mathbf{e}_2\cdot\mathbf{i} = -k\sin\varphi$$

Canceling $\sin\varphi$ from both sides yields $k = d\varphi/ds$, which is as expected. □

Example 2.9. A curve L on a paraboloid S intersects a meridian at an angle φ at a point M, as shown in Fig. 2.22. Obtain the normal curvature k_n, geodesic curvature k_g, and geodesic torsion τ_g of the curve at the point M.

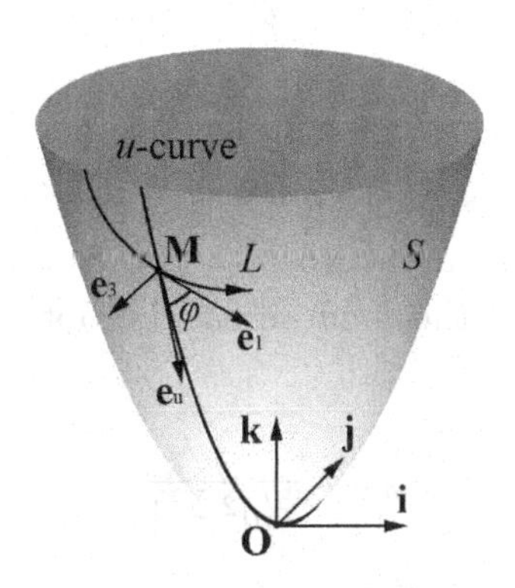

Fig. 2.22: A curve L intersecting a meridian on a paraboloid

Solution:

The paraboloid S can be parameterized as

$$\mathbf{x}(u,v) = \begin{bmatrix} u\cos v & u\sin v & \frac{u^2}{2} \end{bmatrix}$$

The coefficients of the first fundamental forms are, respectively,

$$E = \mathbf{x}_u \cdot \mathbf{x}_u = u^2 + 1$$
$$F = \mathbf{x}_u \cdot \mathbf{x}_v = 0$$
$$G = \mathbf{x}_v \cdot \mathbf{x}_v = u^2$$

The coefficients of the second fundamental forms are, respectively,

$$L = \mathbf{N} \cdot \mathbf{x}_{uu} = \frac{1}{\sqrt{u^2+1}}$$
$$M = \mathbf{N} \cdot \mathbf{x}_{uv} = 0$$
$$N = \mathbf{N} \cdot \mathbf{x}_{vv} = \frac{u^2}{\sqrt{u^2+1}}$$

It follows from Eq. (2.56) that the normal curvature k_{nu}, geodesic curvatures k_{gu}, and geodesic torsion τ_{gu} of the u-curve are, respectively,

$$k_{nu} = \frac{1}{(u^2+1)^{\frac{3}{2}}}$$
$$k_{gu} = 0$$
$$\tau_{gu} = 0$$

Similarly, the normal curvature k_{nv}, geodesic curvatures k_{gv}, and geodesic torsion τ_{gv} of the v-curve can be obtained from Eq. (2.58) respectively as

$$k_{nv} = \frac{1}{\sqrt{u^2+1}}$$
$$k_{gv} = \frac{1}{u\sqrt{u^2+1}}$$
$$\tau_{gv} = 0$$

The curvatures of the curve L can be obtained from Eq. (2.62) as

$$k_n = \frac{\sin^2(\varphi)}{\sqrt{u^2+1}}$$

$$k_g = -\frac{\sin\varphi}{u\sqrt{u^2+1}} + \frac{d\varphi}{ds}$$

$$\tau_g = \frac{u^2 \sin 2\varphi}{2(u^2+1)^{\frac{3}{2}}}$$

□

2.11 Conclusion

We have used the moving-frame method to investigate the geometry of curves and surfaces in 3D Euclidean space. The curvature and torsion of the Frenet frame of a curve reveals the extent of the curve's bending in the osculating plane and twisting about the binormal direction. The geodesic curvature, normal curvature, and geodesic torsion of the Darboux frame of a surface curve reveals the relationship between the surface and the curve. In particular, some characteristic curves, such as line of curvature and geodesics, can be deduced from the values of these curvatures.

Chapter 3

The Adjoint Approach to Curves and Surfaces

In this chapter, we introduce the adjoint approach, which is a powerful tool to study kinematic differential geometry when two geometric entities, such as curves and surfaces are present. The basic idea is to observe the changes of one entity while the observer moves with the moving frame of the other entity.

3.1 The Adjoint Approach to a Planar Curve

Let L denote a planar curve and the Frenet frame $(\mathbf{M}, \mathbf{e}_1, \mathbf{e}_2)$ moves along L. Recall that the torsion of a planar curve L vanishes, so the Frenet formulas of L can be obtained from Eq. (2.24) as

$$\frac{d\mathbf{M}}{ds} = \mathbf{e}_1$$

$$\frac{d}{ds}\begin{bmatrix}\mathbf{e}_1\\ \mathbf{e}_2\end{bmatrix} = \begin{bmatrix}0 & k\\ -k & 0\end{bmatrix}\begin{bmatrix}\mathbf{e}_1\\ \mathbf{e}_2\end{bmatrix}$$

It is understood that the differentiation of $\mathbf{e}_3$ is 0 as $\mathbf{e}_3$, the normal vector of the plane in which L lies, is a constant.

We assume that in the same plane a moving point $\mathbf{P}(s)$ exists corresponding to each $\mathbf{M}(s)$ of L, giving a trajectory L'. The curve L' is deemed an adjoint curve to L, as shown in Fig. 3.1.

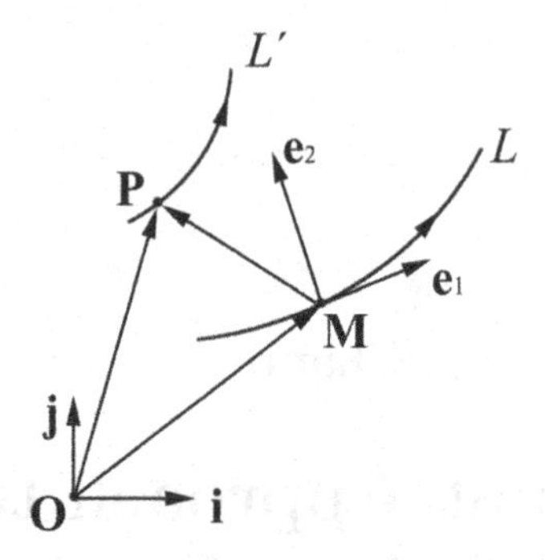

Fig. 3.1: The planar curve L' is adjoint to the planar curve L

Suppose that the coordinates of the point P w.r.t. the frame ($\mathbf{M}$, $\mathbf{e}_1$, $\mathbf{e}_2$) are $(u_1(s), u_2(s))$, where s is the arc length of the curve L. Then the position vector of the point P referenced w.r.t. the frame ($\mathbf{O}$, $\mathbf{i}$, $\mathbf{j}$) is

$$\mathbf{P}(s) = \mathbf{M}(s) + u_1(s)\mathbf{e}_1(s) + u_2(s)\mathbf{e}_2(s) \tag{3.1}$$

Differentiating both sides of Eq. (3.1) w.r.t. s yields

$$\frac{d\mathbf{P}}{ds} = A\mathbf{e}_1 + B\mathbf{e}_2 \tag{3.2}$$

where

$$A = \frac{du_1}{ds} - ku_2 + 1$$

$$B = \frac{du_2}{ds} + ku_1$$

Remark 3.1. Note that (A, B) represent the kinematic velocity of the point P w.r.t. the frame ($\mathbf{O}$, $\mathbf{i}$, $\mathbf{j}$), while $(du_1/ds, du_2/ds)$ represents its kinematic velocity w.r.t. the Frenet frame ($\mathbf{M}$, $\mathbf{e}_1$, $\mathbf{e}_2$), i.e., the velocity observed from the Frenet frame.

Remark 3.2. In the study by Wang and Wang (2015) and Sasaki (1955), a metaphor was used to help understand the adjoint approach. The original curve L can be visualized as a winding river on which a boat is moving. A boatman observes the locus of a moving panther P w.r.t. the frame ($\mathbf{M}$, $\mathbf{e}_1$, $\mathbf{e}_2$) attached to the boat such that the boatman is at the position M, $\mathbf{e}_1$ is in the direction from the back to head, and $\mathbf{e}_2$ is perpendicular to $\mathbf{e}_1$, as shown in Fig. 3.2. The boatman is then able to calculate the geometry of the locus traced by the panther by the measurements of the river and the coordinates.

Example 3.1. Obtain the Frenet formulas at the point P of the curve L' in Eq. (3.1).

Solution:
Set up the Frenet frame $(\mathbf{P},\ \mathbf{E}_1,\ \mathbf{E}_2)$ at the point P, as shown in Fig. 3.3.

Let s' represent the arc length of the curve L'. It follows from Eq. (3.2) that

$$\frac{ds'}{ds} = \left|\frac{d\mathbf{P}}{ds}\right| = \sqrt{A^2 + B^2}$$

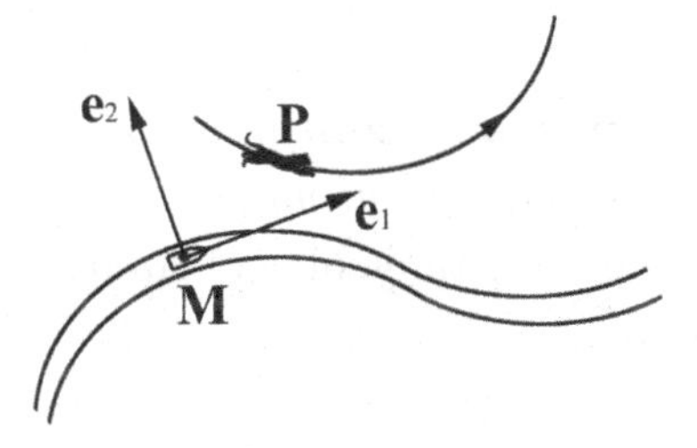

Fig. 3.2: Observing the locus traced by a panther from a moving boat

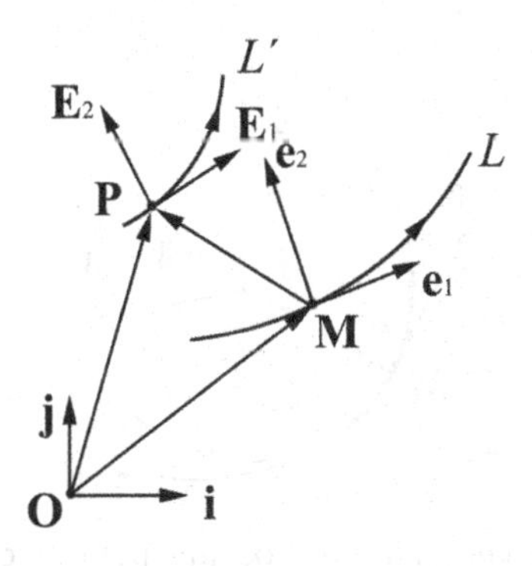

Fig. 3.3: The Frenet frame at the point P on the curve L'

The unit tangent vector $\mathbf{E}_1$ of L' can be obtained as

$$\mathbf{E}_1 = \frac{d\mathbf{P}}{ds'} = \frac{\frac{d\mathbf{P}}{ds}}{\frac{ds'}{ds}} = \frac{A\mathbf{e}_1 + B\mathbf{e}_2}{\sqrt{A^2 + B^2}}$$

The unit normal vector $\mathbf{E}_2$ is perpendicular to $\mathbf{E}_1$. It follows that

$$\mathbf{E}_2 = \frac{-B\mathbf{e}_1 + A\mathbf{e}_2}{\sqrt{A^2 + B^2}}$$

Differentiating the vector $\mathbf{E}_1$ w.r.t. s' yields

$$\frac{d\mathbf{E}_1}{ds'} = \frac{\frac{d\mathbf{E}_1}{ds}}{\frac{ds'}{ds}} = \frac{k(A^2 + B^2) + A\frac{dB}{ds} - B\frac{dA}{ds}}{(A^2 + B^2)^2}(-B\mathbf{e}_1 + A\mathbf{e}_2)$$

The curvature k' of the curve L' can be obtained as

$$k' = \frac{d\mathbf{E}_1}{ds'} \cdot \mathbf{E}_2 = \frac{1}{\sqrt{A^2 + B^2}}\left(k + \frac{A\frac{dB}{ds} - B\frac{dA}{ds}}{A^2 + B^2}\right) \tag{3.3}$$

□

Definition 3.1. Curvature center. The *curvature center* is defined as the center of the *osculating circle* of a curve L at a given point M, which has the same tangent and curvature as L at the point M, as shown in Fig. 3.4.

Example 3.2. Find a curve L such that a straight line bisects each point of L and the curvature center at the corresponding point.

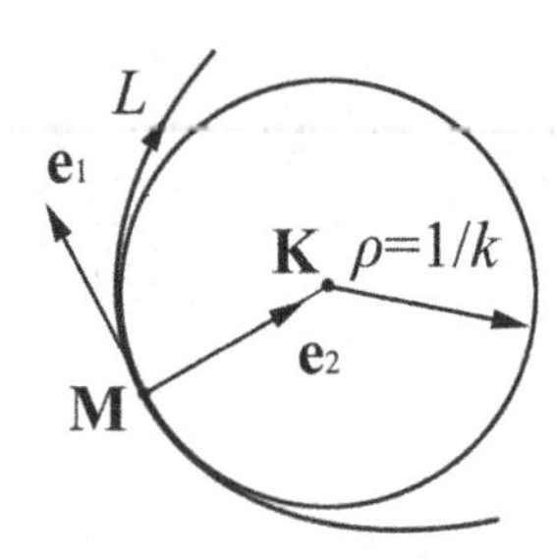

Fig. 3.4: The curvature center K and osculating circle of the curve L at the point M

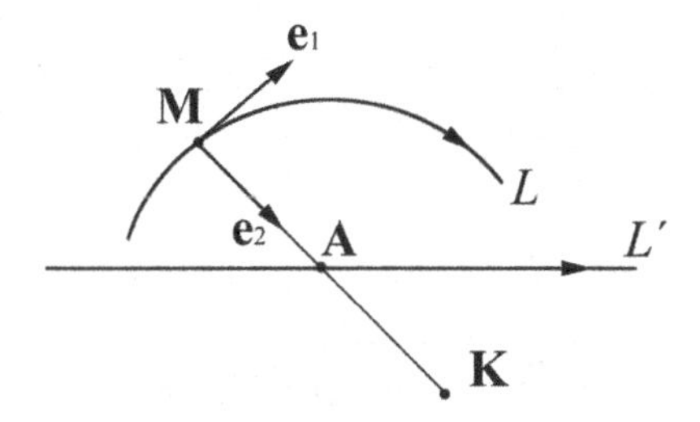

Fig. 3.5: The curve L and its adjoint curve L'

Solution:

Let M denote an arbitrary point on the curve L, and set up the Frenet frame ($\mathbf{M}$, $\mathbf{e}_1$, $\mathbf{e}_2$), as shown in Fig. 3.5.

Let K represent the curvature center of L at M. Then the coordinates of the point K w.r.t. the frame ($\mathbf{M}$, $\mathbf{e}_1$, $\mathbf{e}_2$) is

$$\mathbf{K} = \rho \mathbf{e}_2$$

where $\rho = 1/k$ is the radius of the osculating circle of L at M. Let L' represent the straight line bisecting the line section MK at the point A, then L' is deemed to be the adjoint curve to L. The coordinates of the point A w.r.t. the frame ($\mathbf{M}$, $\mathbf{e}_1$, $\mathbf{e}_2$) are

$$u_1 = 0, \quad u_2 = \frac{\rho}{2}$$

It follows from Eq. (3.2) that

$$A = \frac{1}{2}, \quad B = \frac{1}{2}\frac{d\rho}{ds}$$

where s is the arc length of L. The curve L' is a straight line, implying its curvature is 0. It follows from Eq. (3.3) in Example 3.1 that

$$k + \frac{\frac{d^2\rho}{ds^2}}{1 + \left(\frac{d\rho}{ds}\right)^2} = 0 \tag{3.4}$$

Recall that we have obtained $k = \frac{d\theta}{ds}$ in Example 2.8, where θ is the angle made by $\mathbf{e}_1$ and the $\mathbf{i}$-axis. It follows that

$$\frac{d\theta}{ds} + \frac{d\tan^{-1}(\frac{d\rho}{ds})}{ds} = 0$$

Then

$$\tan^{-1}\left(\frac{d\rho}{ds}\right) = -(\theta + C_1)$$

where C_1 is a constant. Note that $ds = \rho d\theta$, then it follows that

$$\frac{d\rho}{\rho} = -\tan(\rho + C_1)d\theta$$

Then

$$\rho = C_2 \cos(\theta + C_1) \tag{3.5}$$

where C_2 is a constant. Since $ds = \rho d\theta$, it follows that

$$ds = C_2 \cos(\theta + C_1)d\theta$$

Then

$$s = C_2 \sin(\theta + C_1) + C_3 \tag{3.6}$$

where C_3 is a constant. If we let $C_3 = 0$, it follows from Eqs. (3.5) and (3.6) that

$$s^2 + \rho^2 = C_2^2 \tag{3.7}$$

This gives the natural equation of a cycloid, which is traced by a point on a circle rolling along a straight line without slipping, as shown in Fig. 3.6. □

Example 3.3. The *evolute* of a curve is the locus of all its curvature centers. Let L' denote the evolute of a curve L, as shown in Fig. 3.7. Obtain the curvature of L'.

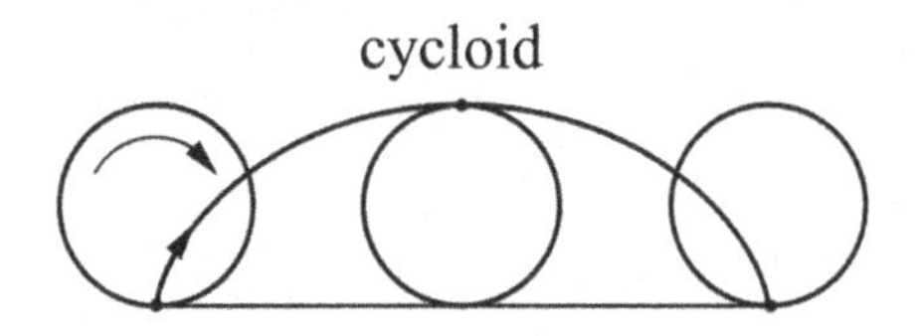

Fig. 3.6: A cycloid by a point on a circle rolling along a straight line

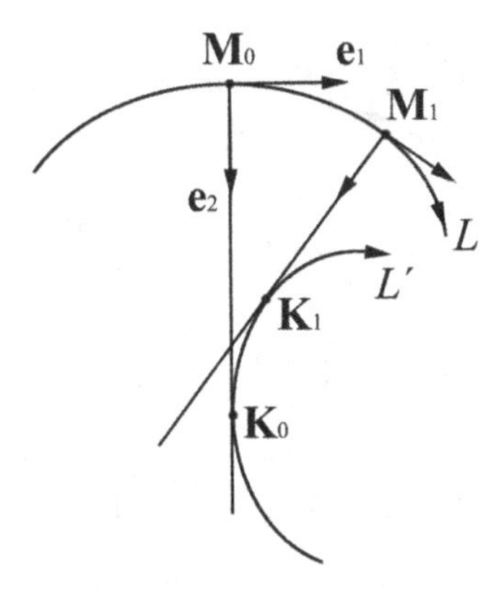

Fig. 3.7: The curve L and its evolute L'

Solution:

Let M denote an arbitrary point on the curve L, and set up the Frenet frame ($\mathbf{M}$, $\mathbf{e}_1$, $\mathbf{e}_2$), as shown in Fig. 3.7. Let K represent the curvature center of L at M. Then the coordinates of the point K w.r.t. the frame ($\mathbf{M}$, $\mathbf{e}_1$, $\mathbf{e}_2$) is

$$\mathbf{K} = \rho \mathbf{e}_2$$

where $\rho = 1/k$ is the radius of the osculating circle of L at M. It is clear that L' is the locus traced by K, and can be deemed as the adjoint curve to L. It follows from Eq. (3.1) that

$$u_1 = 0, \quad u_2 = \rho$$

It follows from Eq. (3.2) that

$$A = 0, \quad B = \frac{d\rho}{ds}$$

Hence, the curvature k' of L' can be obtained from Eq. (3.3) as

$$k' = \frac{k}{\left|\frac{d\rho}{ds}\right|} = \left(\rho \left|\frac{d\rho}{ds}\right|\right)^{-1}$$

□

Definition 3.2. Involute and Evolute. Attach an imaginary taut string to a given curve and wind the string on the given curve.

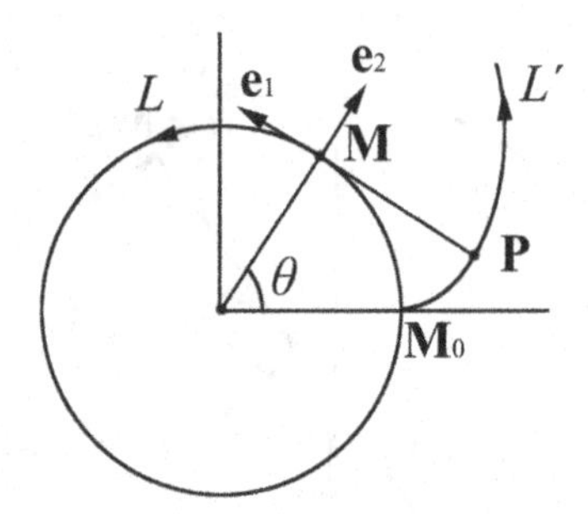

Fig. 3.8: The involute of a circle of radius R

The locus of points traced out by the end of the string is called the involute of the original curve, and the original curve is called the evolute of its involute.

Example 3.4. Find the curvature of the involute L' of the circle L of radius R, starting at $\theta = 0$, as shown in Fig. 3.8.

Solution:
Let M denote an arbitrary point on the circle L, and set up the Frenet frame $(\mathbf{M}, \mathbf{e}_1, \mathbf{e}_2)$. Let s denote arc length of the circle L, then the length of MP is equal to the arc length M_0M from the definition of an involute. It follows that the involute L' can be represented as

$$\mathbf{P} = \mathbf{M} - s\mathbf{e}_1$$

We treat L' as the adjoint curve to L. It follows from Eq. (3.1) that

$$u_1 = -s, \quad u_2 = 0$$

It follows from Eq. (3.2) that

$$A = 0, \quad B = -\frac{s}{R}$$

Hence, the curvature k' of L' can be obtained from Eq. (3.3) as

$$k' = \frac{1}{s} = \frac{1}{R\theta}$$

where θ is the central angle corresponding to the arc M_0M. □

3.2 The Fixed-Point Conditions and Roulettes

In particular, if the point P is fixed w.r.t. the frame $(\mathbf{O}, \mathbf{i}, \mathbf{j})$, as shown in Fig. 3.9, the kinematic velocity of this point referenced w.r.t. the frame $(\mathbf{O}, \mathbf{i}, \mathbf{j})$ must be 0, i.e., $d\mathbf{P}/ds = 0$ in Eq. (3.2), which implies $A = 0, B = 0$.

However, the geometric velocity of the point P is not 0 when viewed from the moving frame $(\mathbf{M}, \mathbf{e}_1, \mathbf{e}_2)$. The coordinates of the point P referenced w.r.t. the frame $(\mathbf{M}, \mathbf{e}_1, \mathbf{e}_2)$ are (u_1, u_2), thus the corresponding geometric velocity is $(du_1/ds, du_2/ds)$, which can be obtained by setting $A = 0, B = 0$ from Eq. (3.2) as

$$\begin{aligned} \frac{du_1}{ds} &= ku_2 - 1 \\ \frac{du_2}{ds} &= -ku_1 \end{aligned} \tag{3.8}$$

Definition 3.3. The Fixed-Point Conditions. Equation (3.8) is called the fixed-point conditions, which gives the geometric velocity of the fixed point P w.r.t. the moving frame $(\mathbf{M}, \mathbf{e}_1, \mathbf{e}_2)$.

Remark 3.3. The same metaphor in Remark 3.2 can be used to visualize the fixed-point conditions. The panther remains static now, so its velocity for an observer on the riverbank is 0. However, its velocity observed from the boat varies as the boat moves along the river.

The fixed-point conditions facilitates the investigation of the geometry of roulettes, which are defined as follows.

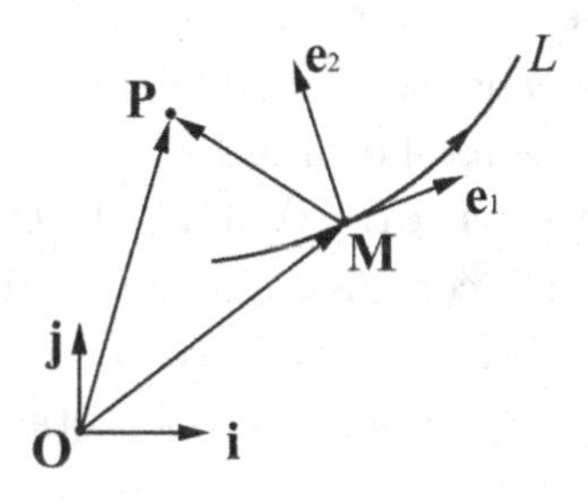

Fig. 3.9: The point P is fixed w.r.t. the curve L

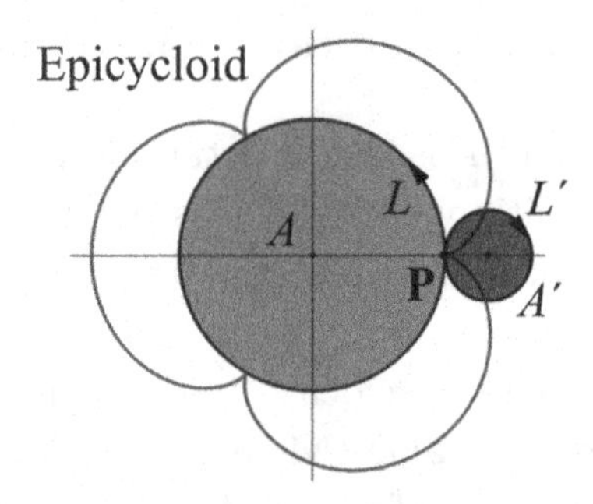

Fig. 3.10: An epicycloid generated by a circle of radius 1 rolling without slipping on a circle of radius 3

Definition 3.4. Roulette. Let a lamina A' rolls without slipping on a lamina A along a curve L' on the lamina A' and a curve L on the lamina A. The curve L_P traced by a point P attached to the lamina A' is called a roulette.

For example, a point fixed on a circle rolling on a straight line generates a cycloid, as shown in Fig. 3.6, and a point fixed on a circle rolling on another circle generates an epicycloid, as shown in Fig. 3.10.

To examine the geometry of the roulette L_P, we set up a fixed frame ($\mathbf{O}$, $\mathbf{i}$, $\mathbf{j}$) on the object A and a fixed frame ($\mathbf{O}'$, $\mathbf{i}'$, $\mathbf{j}'$) on the object A'. Let M denote the contact point on L and M' on L'. We set up the Frenet frame ($\mathbf{M}$, $\mathbf{e}_1$, $\mathbf{e}_2$) on L and ($\mathbf{M}'$, $\mathbf{e}'_1$, $\mathbf{e}'_2$) on L'. The geometric constraints imposed by the rolling contact include that the two curves being tangent at the contact points and the arc lengths covered in the same period of time being equal. Hence, the two contact points M and M' coincide and the two Frenet frames can be made to coincide, as shown in Fig. 3.11, where the frame ($\mathbf{M}$, $\mathbf{e}_1$, $\mathbf{e}_2$) represents both ($\mathbf{M}$, $\mathbf{e}_1$, $\mathbf{e}_2$) and ($\mathbf{M}'$, $\mathbf{e}'_1$, $\mathbf{e}'_2$).

Let (u'_1, u'_2) denote the coordinates of the point P referenced w.r.t. the frame ($\mathbf{M}'$, $\mathbf{e}'_1$, $\mathbf{e}'_2$). The position vector P' referenced w.r.t. the frame ($\mathbf{O}'$, $\mathbf{i}'$, $\mathbf{j}'$) on the lamina A' can be obtained as

$$\mathbf{P}' = \mathbf{M}' + u'_1\mathbf{e}'_1 + u'_2\mathbf{e}'_2 \tag{3.9}$$

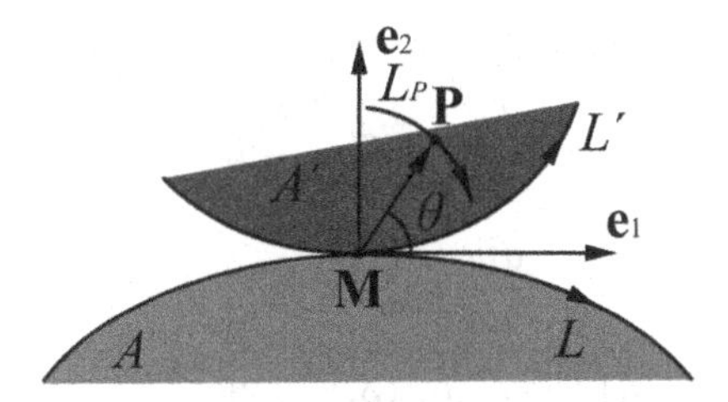

Fig. 3.11: The roulette L_P generated by rolling the curve L on L'

Let s' denote the arc length and k' the curvature of L'. Since the point P is fixed on the lamina A', the fixed-point conditions in Eq. (3.8) give

$$\begin{aligned} \frac{du_1'}{ds'} &= k'u_2' - 1 \\ \frac{du_2'}{ds'} &= -k'u_1' \end{aligned} \tag{3.10}$$

Similarly, let (u_1, u_2) denote the coordinates of the point P referenced w.r.t. the Frenet frame $(\mathbf{M}, \mathbf{e}_1, \mathbf{e}_2)$, s the arc length of L, and k the curvature of L. The position vector for the point P referenced w.r.t. the frame $(\mathbf{O}, \mathbf{i}, \mathbf{j})$ on the object A can be obtained as

$$\mathbf{P} = \mathbf{M} + u_1\mathbf{e}_1 + u_2\mathbf{e}_2 \tag{3.11}$$

The geometric velocity of the point P can be obtained by differentiating both sides of Eq. (3.11):

$$\frac{d\mathbf{P}}{ds} = \left(\frac{du_1}{ds} - ku_2 + 1\right)\mathbf{e}_1 + \left(\frac{du_2}{ds} + ku_1\right)\mathbf{e}_2 \tag{3.12}$$

Since we have made the two Frenet frames coincide, it follows that $u_1 = u_1', u_2 = u_2'$. Further, the arc lengths s and s' are equal, which is imposed by the rolling constraints. It thus follows that

$$\frac{du_1}{ds} = \frac{du_1'}{ds'}, \quad \frac{du_2}{ds} = \frac{du_2'}{ds'} \tag{3.13}$$

It follows from Eq. (3.10) that

$$\begin{aligned}\frac{du_1}{ds} &= k'u_2 - 1\\ \frac{du_2}{ds} &= -k'u_1\end{aligned} \tag{3.14}$$

Substituting Eq. (3.14) into Eq. (3.12) yields

$$\frac{d\mathbf{P}}{ds} = (k' - k)\,(u_2\mathbf{e}_1 - u_1\mathbf{e}_2) \tag{3.15}$$

It is clear that the lamina A' rotates about the point M instantaneously, so the curve L is the fixed centrodes and L' the moving centrodes. We are now in the position to examine some famous conclusions in planar kinematics.

Example 3.5. Derive the curvature of the curve L_P.

Solution:

The curve L_P is an adjoint curve to L. It follows from Eqs. (3.2) and (3.15) that

$$\begin{aligned}A &= (k' - k)u_2\\ B &= -(k' - k)u_1\end{aligned}$$

The curvature k_P of L_P can be obtained from Eqs. (3.3) and (3.14) as

$$\begin{aligned}k_P &= \frac{1}{|k' - k|\sqrt{u_1^2 + u_2^2}}\left(k + \frac{-\frac{du_1}{ds}u_2 + \frac{du_2}{ds}u_1}{u_1^2 + u_2^2}\right)\\ &= \frac{1}{|k' - k|\sqrt{u_1^2 + u_2^2}}\left(k - k' + \frac{u_2}{u_1^2 + u_2^2}\right)\end{aligned} \tag{3.16}$$

Let r denote the length of the vector $\mathbf{MP}$, θ the angle between $\mathbf{MP}$ and $\mathbf{e}_1$, and $\epsilon = \text{sgn}(k - k')$. It is clear that $r = \sqrt{u_1^2 + u_2^2}$ and $\sin\theta = u_2/\sqrt{u_1^2 + u_2^2}$. Equation (3.16) can be rewritten as

$$k_P = \epsilon\left(\frac{1}{r} + \frac{\sin\theta}{(k - k')r^2}\right) \tag{3.17}$$

□

Remark 3.4. Let ρ, ρ', ρ_P represent the radius of osculating circles of L, L', and L_P respectively. Equation (3.17) can be straightforwardly reformatted as

$$\frac{1}{\epsilon\rho_P - r} + \frac{1}{r} = \left(\frac{1}{\rho'} - \frac{1}{\rho}\right)\frac{1}{\sin\theta} \tag{3.18}$$

Equation (3.18) is the famous *Euler–Savary* equation, which relates the curvature of an adjoint curve L_P to the fixed and moving centrodes.

Example 3.6. An inflection point is a point on a curve at which the curvature is 0. Prove that all points on the lamina A' that are inflections points are located on a circle, which is called the *inflection circle*.

Solution:
A point on L_p is an inflection point means $k_P = 0$. It follows from Eq. (3.17) that

$$r = \frac{1}{k' - k}\sin\theta = R\sin\theta \tag{3.19}$$

where $R = 1/(k' - k)$. Note that Eq. (3.19) represents a circle, as shown in Fig. 3.12. □

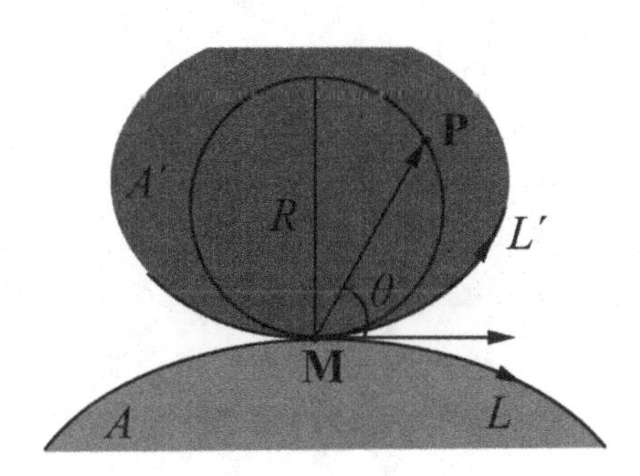

Fig. 3.12: The inflection circle at one instant

3.3 The Adjoint Approach to a Spatial Curve

Let L denote a curve and the Frenet frame ($\mathbf{M}$, $\mathbf{e}_1$, $\mathbf{e}_2$, $\mathbf{e}_3$) moves along L. Let L' denote another curve and P a point on L'. Both curves are referenced w.r.t. a fixed frame ($\mathbf{O}$, $\mathbf{i}$, $\mathbf{j}$, $\mathbf{k}$).

We observe the coordinates of the point P w.r.t. the Frenet frame ($\mathbf{M}$, $\mathbf{e}_1$, $\mathbf{e}_2$, $\mathbf{e}_3$). The curve L' is designated as an adjoint curve to the curve L, as shown in Fig. 3.13.

Remark 3.5. We use another metaphor to better understand this approach. The curve L can be visualize as a roller-coaster track and L' as an adjacent roller-coaster track (suppose there are two roller coasters in the theme park). You ride on the track L in an open car, which corresponds to the point M, and your friend rides on the other track L' in an open car, which corresponds to the point P. You set up a frame ($\mathbf{M}$, $\mathbf{e}_1$, $\mathbf{e}_2$, $\mathbf{e}_3$) fixed on your car in such a way that $\mathbf{e}_1$ is tangent to L and $\mathbf{e}_2$ is parallel with the direction of the centrifugal force. When you and your friend are taking the ride simultaneously, you can obtain the coordinates of your position w.r.t. the frame ($\mathbf{O}$, $\mathbf{i}$, $\mathbf{j}$, $\mathbf{k}$) fixed on the ground and obtain the coordinates of your friend's position w.r.t. the frame ($\mathbf{M}$, $\mathbf{e}_1$, $\mathbf{e}_2$, $\mathbf{e}_3$). In this way, you obtain the coordinates of your friend's position w.r.t. the ground frame ($\mathbf{O}$, $\mathbf{i}$, $\mathbf{j}$, $\mathbf{k}$).

Let (u_1, u_2, u_3) denote the coordinates of the point P referenced w.r.t. the frame ($\mathbf{M}$, $\mathbf{e}_1$, $\mathbf{e}_2$, $\mathbf{e}_3$). Then the position vector of the point

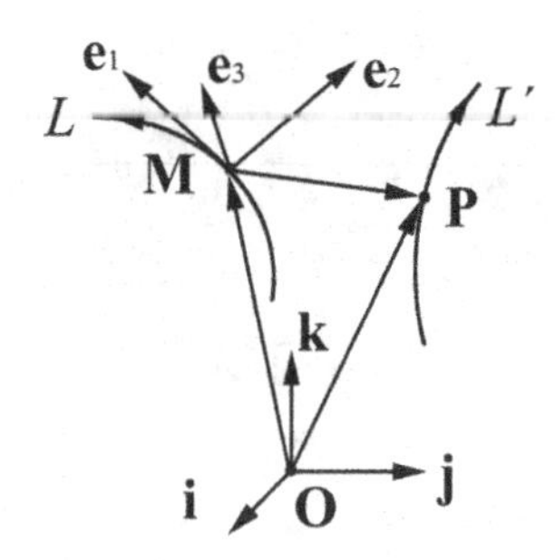

Fig. 3.13: The curve L' is adjoint to the curve L

P referenced w.r.t. the frame ($\mathbf{O}$, $\mathbf{i}$, $\mathbf{j}$, $\mathbf{k}$) can be written as

$$\mathbf{P} = \mathbf{M} + u_1\mathbf{e}_1 + u_2\mathbf{e}_2 + u_3\mathbf{e}_3 \tag{3.20}$$

The geometric velocity of the point P in terms of the arc length s of the curve L can be obtained by differentiating both sides of Eq. (3.20) w.r.t. s as

$$\begin{aligned}\frac{d\mathbf{P}}{ds} &= \mathbf{e}_1 + \frac{du_1}{ds}\mathbf{e}_1 + u_1k\mathbf{e}_2 \\ &\quad + \frac{du_2}{ds}\mathbf{e}_2 + u_2(-k\mathbf{e}_1 + \tau\mathbf{e}_3) \\ &\quad + \frac{du_3}{ds}\mathbf{e}_3 - u_3\tau\mathbf{e}_2 \\ &= \left(\frac{du_1}{ds} + 1 - u_2k\right)\mathbf{e}_1 \\ &\quad + \left(\frac{du_2}{ds} + u_1k - u_3\tau\right)\mathbf{e}_2 \\ &\quad + \left(\frac{du_3}{ds} + u_2\tau\right)\mathbf{e}_3\end{aligned} \tag{3.21}$$

where k is the curvature and τ the torsion of L. The left side of Eq. (3.21) gives the geometric velocity of the point P parameterized by the arc length s w.r.t to the frame ($\mathbf{O}$, $\mathbf{i}$, $\mathbf{j}$, $\mathbf{k}$); the right side of Eq. (3.21) referenced the same velocity w.r.t. the frame ($\mathbf{M}$, $\mathbf{e}_1$, $\mathbf{e}_2$, $\mathbf{e}_3$).

In particular, if the point P is fixed w.r.t. the frame ($\mathbf{O}$, $\mathbf{i}$, $\mathbf{j}$, $\mathbf{k}$), its velocity must be 0, i.e., $d\mathbf{P}/ds = 0$. But this velocity is not 0 when viewed from the moving frame ($\mathbf{M}$, $\mathbf{e}_1$, $\mathbf{e}_2$, $\mathbf{e}_3$), as shown in Fig. 3.14.

The geometric velocity of the point P w.r.t. the frame ($\mathbf{M}$, $\mathbf{e}_1$, $\mathbf{e}_2$, $\mathbf{e}_3$) — (du_1/ds, du_2/ds, du_3/ds) — can be obtained by setting the right side of Eq. (3.21) to 0 as

$$\begin{aligned}\frac{du_1}{ds} &= -1 + u_2k \\ \frac{du_2}{ds} &= -u_1k + u_3\tau \\ \frac{du_3}{ds} &= -u_2\tau\end{aligned} \tag{3.22}$$

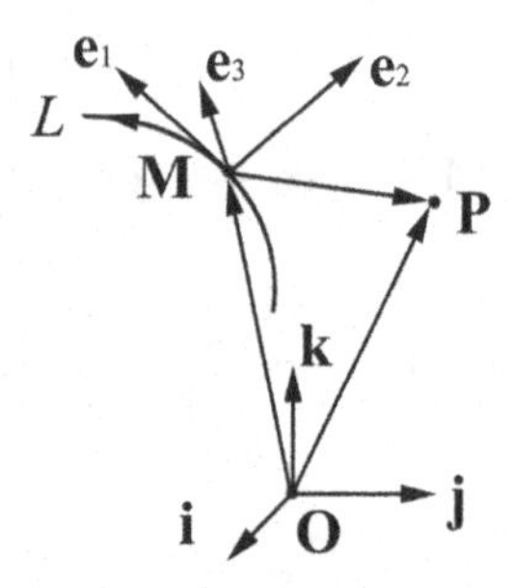

Fig. 3.14: The fixed-point conditions

Similar to Definition 3.3, Eq. (3.22) is called the *fixed-point conditions* for a spatial curve, which give the geometric velocity of the fixed point P referenced w.r.t. the moving frame ($\mathbf{M}$, $\mathbf{e}_1$, $\mathbf{e}_2$, $\mathbf{e}_3$).

Remark 3.6. We use the same roller-coaster metaphor. The point P being fixed can be visualized as your friend's car does not move at all while your are riding. Then the velocity of your friend's car is 0 w.r.t. the ground frame ($\mathbf{O}$, $\mathbf{i}$, $\mathbf{j}$, $\mathbf{k}$), but its velocity w.r.t. your car's frame ($\mathbf{M}$, $\mathbf{e}_1$, $\mathbf{e}_2$, $\mathbf{e}_3$) is not 0, and its velocity components are expressed in Eq. (3.22).

Example 3.7. Two curves L and L' are called *Bertrand curves* if they have common principal normal lines, as shown in Fig. 3.15. Show that on the two curves the distance between corresponding points is constant and the angle between corresponding tangent lines is constant.

Solution:
Let M denote a point on L and P its corresponding point on L' where the two curves have common principal normal lines. Set up the Frenet frame ($\mathbf{M}$, $\mathbf{e}_1$, $\mathbf{e}_2$, $\mathbf{e}_3$) on L and the Frenet frame ($\mathbf{P}$, $\mathbf{E}_1$, $\mathbf{E}_2$, $\mathbf{E}_3$) on L'. It follows that $\mathbf{e}_2$ and $\mathbf{E}_2$ are collinear, as shown in Fig. 3.15.

Let L be parameterized by $\mathbf{x} = \mathbf{x}(s)$, where s is its arc length and let (u_1, u_2, u_3) denote the coordinates of the point P referenced w.r.t. the frame ($\mathbf{M}$, $\mathbf{e}_1$, $\mathbf{e}_2$, $\mathbf{e}_3$). Since u_1 and u_3 are 0, the position

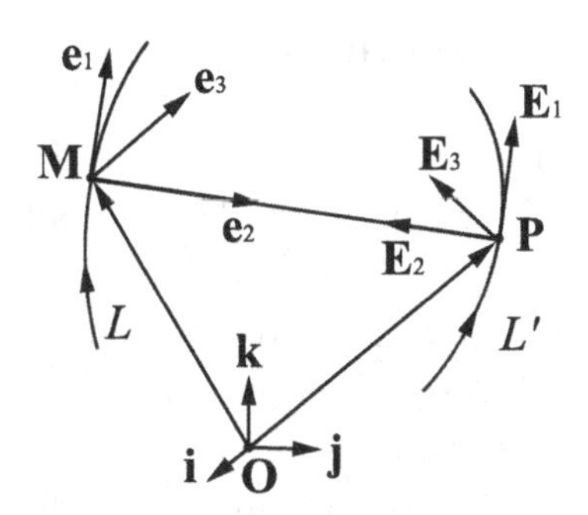

Fig. 3.15: The Bertrand curves

vector of the point P can be expressed as

$$\mathbf{P} = \mathbf{M} + u_2\mathbf{e}_2 \tag{3.23}$$

Differentiating both sides of Eq. (3.23) w.r.t. s yields

$$\begin{aligned}\frac{d\mathbf{P}}{ds} &= \mathbf{e}_1 + \frac{du_2}{ds}\mathbf{e}_2 + u_2(-k\mathbf{e}_1 + \tau\mathbf{e}_3) \\ &= (1 - u_2k)\mathbf{e}_1 + \frac{du_2}{ds}\mathbf{e}_2 + u_2\tau\mathbf{e}_3\end{aligned} \tag{3.24}$$

where k is the curvature and τ the torsion of L.

Note that $d\mathbf{P}/ds$ is tangent to the curve L' and thus orthogonal to $\mathbf{E}_2$ and $\mathbf{e}_2$. Taking the dot product of both sides of Eq. (3.24) with $\mathbf{e}_2$ yields

$$\frac{du_2}{ds} = 0$$

Hence, we have shown the distance between the corresponding points M and P, which is $|u_2|$, is a constant.

Next we show the angle between corresponding tangent lines is constant. For this purpose, we have

$$\frac{d}{ds}(\mathbf{e}_1 \cdot \mathbf{E}_1) = k\mathbf{e}_2 \cdot \mathbf{E}_1 + \mathbf{e}_1 \cdot K\mathbf{E}_2\frac{ds'}{ds} \tag{3.25}$$

where s' is the arc length and K the curvature of L'. It is obvious that the right side of Eq. (3.25) is equal to 0. □

3.4 The Adjoint Approach to a Surface Curve

Let L represent a curve on a surface S and P an arbitrary point on the surface S, as shown in Fig. 3.16. We obtain the fixed-point conditions.

Let M denote a point on L, and we set up the Darboux frame $(\mathbf{M}, \mathbf{e}_1, \mathbf{e}_2, \mathbf{e}_3)$ such that $\mathbf{e}_1$ is the unit tangent and $\mathbf{e}_3$ the surface normal, as shown in Fig. 3.16.

Let (u_1, u_2, u_3) denote the coordinates of the point P w.r.t. the frame $(\mathbf{M}, \mathbf{e}_1, \mathbf{e}_2, \mathbf{e}_3)$. Then the position vector of P w.r.t. a frame fixed on the surface S can be obtained as

$$\mathbf{P} = \mathbf{M} + u_1\mathbf{e}_1 + u_2\mathbf{e}_2 + u_3\mathbf{e}_3 \tag{3.26}$$

Differentiating both sides of Eq. (3.26) w.r.t. arc length s of L yields

$$\begin{aligned}
\frac{d\mathbf{P}}{ds} &= \mathbf{e}_1 + \frac{du_1}{ds}\mathbf{e}_1 + u_1(k_g\mathbf{e}_2 + k_n\mathbf{e}_3) \\
&\quad + \frac{du_2}{ds}\mathbf{e}_2 + u_2(-k_g\mathbf{e}_1 + \tau_g\mathbf{e}_3) \\
&\quad + \frac{du_3}{ds}\mathbf{e}_3 + u_3(-k_n\mathbf{e}_1 - \tau_g\mathbf{e}_2) \\
&= \left(\frac{du_1}{ds} - u_2k_g - u_3k_n + 1\right)\mathbf{e}_1 \\
&\quad + \left(\frac{du_2}{ds} + u_1k_g - u_3\tau_g\right)\mathbf{e}_2 \\
&\quad + \left(\frac{du_3}{ds} + u_1k_n + u_2\tau_g\right)\mathbf{e}_3
\end{aligned} \tag{3.27}$$

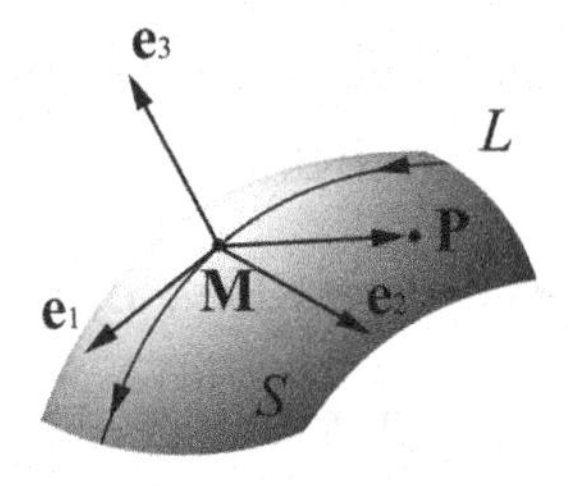

Fig. 3.16: The fixed-point conditions on a surface

where k_n is the normal curvature, k_g the geodesic curvature, and τ_g the geodesic torsion of L. The fixed-point conditions can be obtained by setting the right side of Eq. (3.27) to 0 as

$$\begin{aligned} \frac{du_1}{ds} &= u_2 k_g + u_3 k_n - 1 \\ \frac{du_2}{ds} &= -u_1 k_g + u_3 \tau_g \\ \frac{du_3}{ds} &= -u_1 k_n - u_2 \tau_g \end{aligned} \tag{3.28}$$

Equation (3.28) will be used in the subsequent chapters in this book.

3.5 A Circular Surface with a Fixed Radius: The Surface Adjoint to a Curve

3.5.1 *Definition of a Circular Surface*

Let L denote a space curve and R the radius of a circle. A circular surface is defined to be the surface that is swept out by the circle moving along the space curve L (Cui *et al.*, 2009). Each circle is called a generating circle, which lines in a plane named circle plane, and the curve L acts as the spine curve, as shown in Fig. 3.17.

We apply the adjoint approach to investigating the properties of a circular surface. For this purpose, let $\mathbf{e}_1$ denote the unit normal vector of the circle plane at each point of the curve L, then the

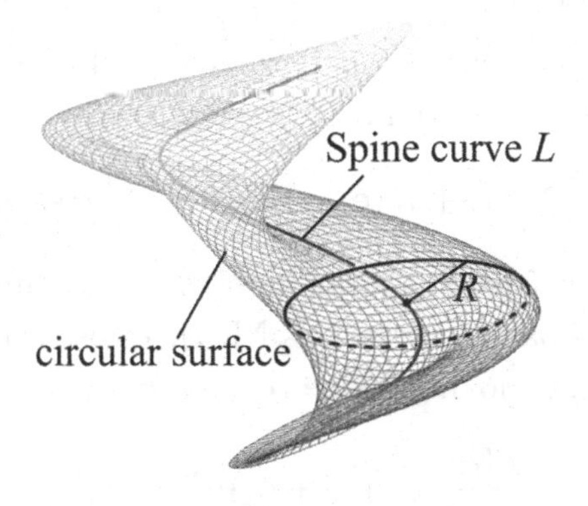

Fig. 3.17: A circular surface with a fixed radius R

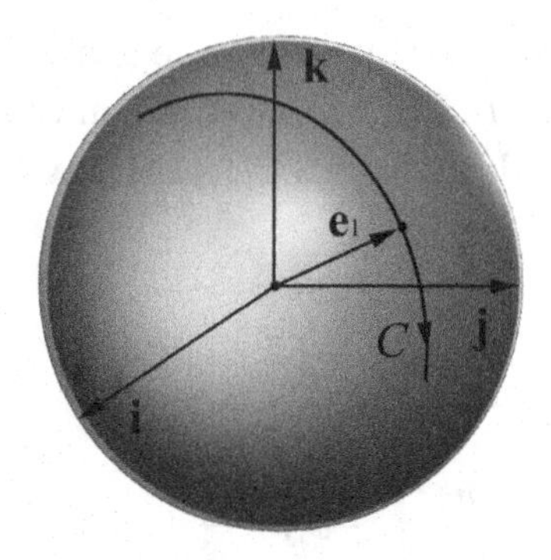

Fig. 3.18: The spherical indicatrix C of the normal $\mathbf{e}_1$ of the circle planes

Gauss map of $\mathbf{e}_1$ produces a spherical indicatrix C on a unit sphere (Lipschutz, 1969), as shown in Fig. 3.18.

Let u represent the arc length of the spherical indicatrix C and $\mathbf{e}_2$ the differentiation of $\mathbf{e}_1(u)$ w.r.t. u, and further let $\mathbf{e}_3 = \mathbf{e}_1 \times \mathbf{e}_2$. It follows that $\mathbf{e}_1 \cdot \mathbf{e}_2 = 0$ and $|\mathbf{e}_2| = 1$, and thus $\mathbf{e}_1$, $\mathbf{e}_2$, and $\mathbf{e}_3$ form a moving frame. It follows from Eq. (2.46) that

$$\frac{d}{du}\begin{bmatrix}\mathbf{e}_1\\ \mathbf{e}_2\\ \mathbf{e}_3\end{bmatrix} = \begin{bmatrix}0 & 1 & 0\\ -1 & 0 & k_g\\ 0 & -k_g & 0\end{bmatrix}\begin{bmatrix}\mathbf{e}_1\\ \mathbf{e}_2\\ \mathbf{e}_3\end{bmatrix} \tag{3.29}$$

where k_g is the geodesic curvature of the spherical indicatrix C.

The moving frame ($\mathbf{e}_1$, $\mathbf{e}_2$, $\mathbf{e}_3$) is attached to each point M of the spine curve L and the vectors $\mathbf{e}_2$ and $\mathbf{e}_3$ span the circle plane at M, as shown in Fig. 3.19.

Suppose the spine curve L is parameterized by u, then a circular surface S can be represented as

$$\mathbf{x}(u,\theta) = \mathbf{M}(u) + R\,(\mathbf{e}_2(u)\cos\theta + \mathbf{e}_3(u)\sin\theta) \tag{3.30}$$

The circular surface S is adjoint to the spine curve L. Note that the spine curve is parameterized as $\mathbf{M}(u)$, whose tangent vector can be expressed in the moving frame ($\mathbf{e}_1$, $\mathbf{e}_2$, $\mathbf{e}_3$) as

$$\frac{d\mathbf{M}}{du} = \alpha_1\mathbf{e}_1 + \alpha_2\mathbf{e}_2 + \alpha_3\mathbf{e}_3 \tag{3.31}$$

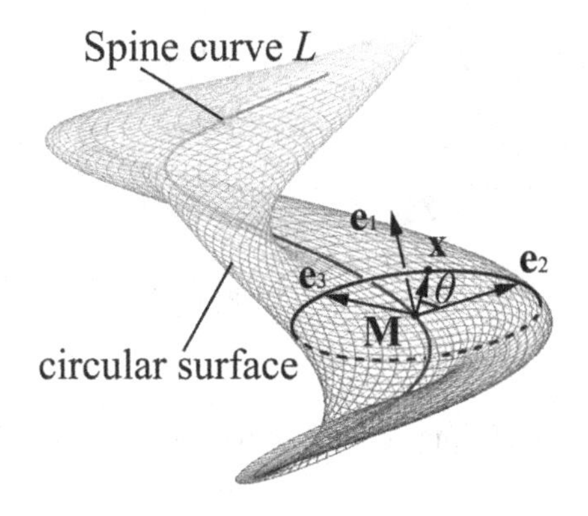

Fig. 3.19: A cross-section of a circular surface

where the scalar variables α_i are the projection of the tangent vector $d\mathbf{M}/du$ onto the $\mathbf{e}_i$-axes ($i = 1, 2, 3$).

The four scalar variables k_g and α_i, $i = 1, 2, 3$, form a complete system of Euclidean invariants of a circular surface. Further, these invariants can be used to construct a circular surface with a fixed radius R through the following equation:

$$\mathbf{M}(u) = \mathbf{M}_0 + \int_{u_0}^{u} (\alpha_1 \mathbf{e}_1 + \alpha_2 \mathbf{e}_2 + \alpha_3 \mathbf{e}_3) du \tag{3.32}$$

These four invariants reveal the properties and facilitate the study of a circular surface. This can be demonstrated by the following sections on constraint circular surfaces, which are workspace surfaces generated by spatial linkage mechanisms. For example, Fig. 3.20 shows such a circular surface generated by a point P on the end-effector of a serially connected RR-linkage mechanism, where R denotes the revolute joint.

We use the set of four Euclidean variants to prove the necessary and sufficient conditions for a circular surface to be one of the workspace surfaces that are generated by serially-connected $C'R$, HR, and RR linkage mechanisms, where H and R represent helical and revolute joints, respectively, and C' represents a 1-DOF special helical joint with a variable pitch.

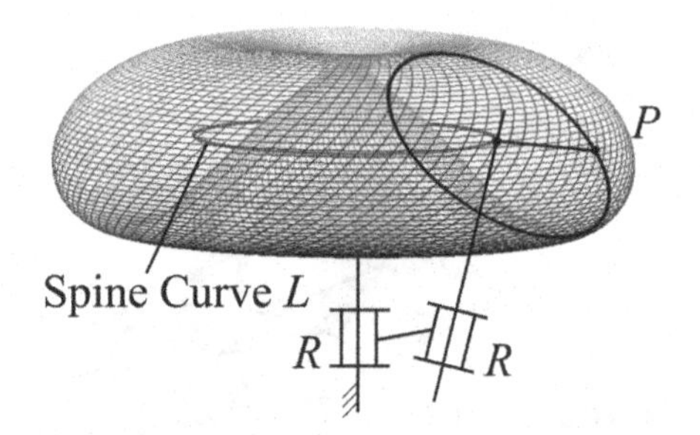

Fig. 3.20: A workspace surface generated by a RR-linkage mechanism

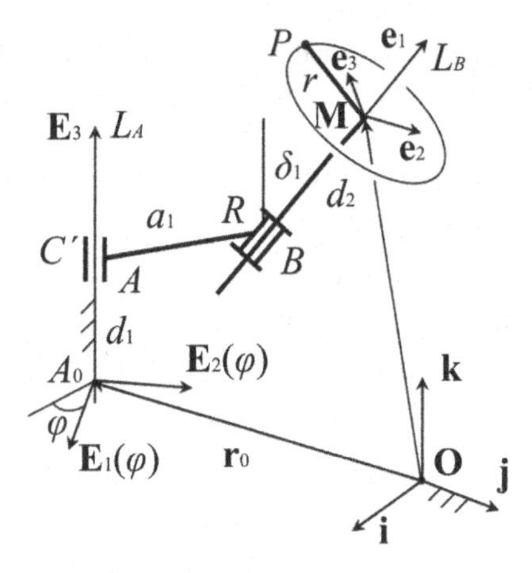

Fig. 3.21: A serially connected $C'R$ linkage mechanism

3.5.2 *The Necessary and Sufficient Conditions for a Circular Surface to be a $C'R$ Workspace Surface*

A $C'R$ workspace surface can be generated by a point P on the end-effector of a serially connected $C'R$ linkage mechanism, as shown in Fig. 3.21.

Let L_A denote the fixed line aligned with the C'-joint axis and L_B the moving line aligned with the R-joint axis, which makes an angle δ with L_A. Let AB denote the common perpendicular line of L_A and L_B, and a_1 the length of AB. Let P denote a point on the end-effector attached to the line L_B, M a point such that PM is perpendicular to L_B, and d_2 the distance between the point M and B.

It is clear that the curve generated by the point M serves as the spine curve for the constraint circular surface generated by the point P. The position vector of the point M referenced w.r.t. a fixed frame can be obtained as

$$\mathbf{M}(\varphi) = \mathbf{r}_0 + d_1\mathbf{E}_3 + a_1\frac{\mathbf{E}_3 \times \mathbf{e}_1(\varphi)}{|\mathbf{E}_3 \times \mathbf{e}_1(\varphi)|} + d_2\mathbf{e}_1(\varphi) \tag{3.33}$$

where $\mathbf{E}_3$ represents the direction of L_A, $\mathbf{e}_1$ the direction of L_B, $\mathbf{E}_3 \times \mathbf{e}_1(\varphi)$ the direction of AB, and φ the rotational angle of the C' joint.

Let $\mathbf{E}_1(\varphi)$ denote unit vector in the direction of the projection of $\mathbf{e}_1$ on the plane perpendicular to $\mathbf{E}_3$, and $\mathbf{E}_2 = \mathbf{E}_3 \times \mathbf{E}_2$, as shown in Fig. 3.22.

The vectors $\mathbf{E}_1$, $\mathbf{E}_2$, $\mathbf{E}_3$ form an orthonormal frame and $\mathbf{E}_1$ and $\mathbf{E}_2$ are related by

$$\frac{d\mathbf{E}_1(\varphi)}{d\varphi} = \mathbf{E}_2(\varphi), \quad \frac{d\mathbf{E}_2(\varphi)}{d\varphi} = -\mathbf{E}_1(\varphi) \tag{3.34}$$

Then the unit vector $\mathbf{e}_1$ can be obtained as

$$\mathbf{e}_1(\varphi) = \cos\delta_1\mathbf{E}_3 + \sin\delta_1\mathbf{E}_1(\varphi) \tag{3.35}$$

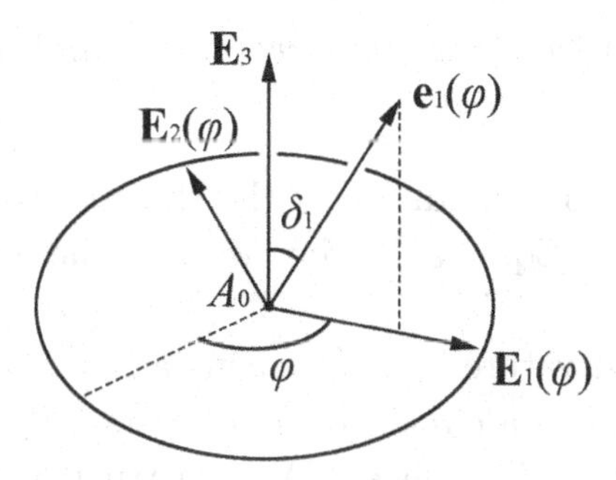

Fig. 3.22: The relationship between $\mathbf{E}_3$, $\mathbf{E}_1(\varphi)$, and $\mathbf{e}_1(\varphi)$

Let u denote the arc length of the spherical indicatrix of $\mathbf{e}_1$. It follows from Eq. (3.35) that

$$du = |d\mathbf{e}_1| = \sin\delta_1 d\varphi \tag{3.36}$$

The vectors $\mathbf{e}_2$ and $\mathbf{e}_3$ that span the circle plane at the point M can be obtained as

$$\begin{aligned} \mathbf{e}_2(\varphi) &= \frac{d\mathbf{e}_1}{du} = \frac{d\mathbf{e}_1}{d\varphi}\frac{d\varphi}{du} = \mathbf{E}_2 \\ \mathbf{e}_3(\varphi) &= \mathbf{e}_1 \times \mathbf{e}_2 = \sin\delta_1\mathbf{E}_3 - \cos\delta_1\mathbf{E}_1(\varphi) \end{aligned} \tag{3.37}$$

The Euclidean invariants k_g in Eq. (3.29) and α_i in Eq. (3.31) can be obtained as

$$\begin{aligned} k_g &= \frac{d\mathbf{e}_2}{du} \cdot \mathbf{e}_3 = \cot\delta_1 \\ \alpha_1 &= \frac{d\mathbf{M}}{du} \cdot \mathbf{e}_1 = \frac{dd_1}{du}\cos\delta_1 - a_1 \\ \alpha_2 &= \frac{d\mathbf{M}}{du} \cdot \mathbf{e}_2 = d_2 \\ \alpha_3 &= \frac{d\mathbf{M}}{du} \cdot \mathbf{e}_3 = \frac{dd_1}{du}\sin\delta_1 + a_1\cot\delta_1 \end{aligned} \tag{3.38}$$

We are now in the position to present the necessary and sufficient conditions for a circular surface to be a $C'R$ surface.

Theorem 3.1. *A circular surface S is a $C'R$ workspace surface if and only if: (1) the geodesic curvature k_g and α_2 are constants; (2) $\alpha_1 - \alpha_3 k_g$ is a constant.*

Proof. The necessary condition of the theorem can be straightforwardly verified from Eq. (3.38). We now prove the sufficient condition in the following.

The geodesic curvature k_g being a constant implies that the spherical indicatrix of $\mathbf{e}_1$ is a circle, so there exists a fixed direction that makes a constant angle with $\mathbf{e}_1$. We construct a coordinate frame $(\mathbf{O}, \mathbf{i}, \mathbf{j}, \mathbf{k})$ such that the $\mathbf{k}$-axis is aligned with the fixed direction, as shown in Fig. 3.23.

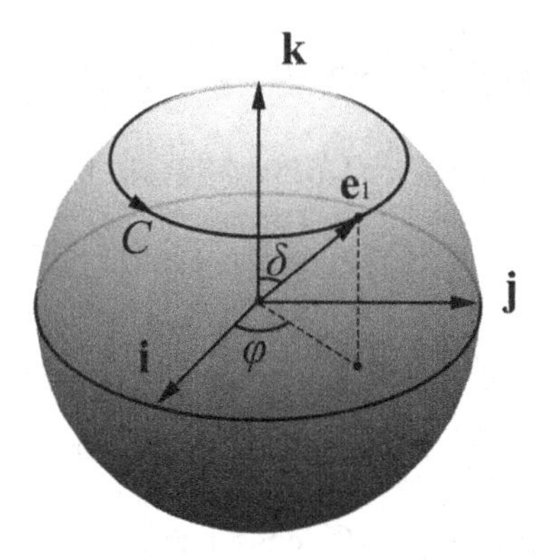

Fig. 3.23: The spherical indicatrix of $\mathbf{e}_1$ being a circle

The vectors $\mathbf{e}_1$, $\mathbf{e}_2$, and $\mathbf{e}_3$ can be referenced w.r.t. the frame ($\mathbf{O}$, $\mathbf{i}$, $\mathbf{j}$, $\mathbf{k}$) as

$$\begin{aligned}
\mathbf{e}_1 &= [\sin\delta\cos\varphi \;\; \sin\delta\sin\varphi \;\; \cos\delta] \\
\mathbf{e}_2 &= [-\sin\varphi \;\; \cos\varphi \;\; 0] \\
\mathbf{e}_3 &= [-\cos\delta\cos\varphi \;\; -\cos\delta\sin\varphi \;\; \sin\delta]
\end{aligned} \tag{3.39}$$

where δ denote the constant angle between $\mathbf{e}_1$ and $\mathbf{k}$ and φ the angle between the projection of $\mathbf{e}_1$ on the plane spanned by the $\mathbf{i}$ and $\mathbf{j}$ axes.

The differentiation of the arc length u of the spherical indicatrix of $\mathbf{e}_1$ can be obtained as

$$du = |d\mathbf{e}_1| = \sin\delta d\varphi \tag{3.40}$$

Let $\mathbf{M}$ denote the spine curve L and suppose that $\alpha_1 - \alpha_3 k_g = C_0$, where C_0 is a constant. The three components of $\mathbf{M}$ can be obtained from Eq. (3.32) as

$$\begin{aligned}
M_x = M_{x0} + \int_0^u &((C_0 + \alpha_3\cot\delta)\sin\delta\cos\varphi - \alpha_2\sin\varphi \\
&- \alpha_3\cos\delta\cos\varphi)du
\end{aligned}$$

$$
\begin{aligned}
&= M_{x0} + \int_0^{\varphi} (C_0 \sin\delta \cos\varphi - \alpha_2 \sin\varphi) \sin\delta d\varphi \\
&= \sin\delta(\alpha_2 \cos\varphi + C_0 \cos\delta \sin\varphi) + C_1 \\
M_y &= M_{y0} + \int_0^{u} ((C_0 + \alpha_3 \cot\delta) \sin\delta \sin\varphi + \alpha_2 \cos\varphi \\
&\quad - \alpha_3 \cos\delta \sin\varphi) du \\
&= M_{y0} + \int_0^{\varphi} (C_0 \sin\delta \sin\varphi + \alpha_2 \cos\varphi) \sin\delta d\varphi \\
&= \sin\delta(\alpha_2 \sin\varphi - C_0 \cos\delta \cos\varphi) + C_2 \\
M_z &= M_{z0} + \int_0^{u} ((C_0 + \alpha_3 \cot\delta) \cos\delta + \alpha_3 \sin\delta)\, du \qquad (3.41)
\end{aligned}
$$

where C_1 and C_2 are constants. We choose the start point of the spine curve such that $C_1 = C_2 = 0$. The distance d between the spine curve and the **k**-axis is

$$
d = |\mathbf{M} \times \mathbf{k}| = \sqrt{\sin^2\delta(\alpha_2^2 + C_0^2 \cos^2\delta)} \qquad (3.42)
$$

It can be seen that the distance d in Eq. (3.42) is a constant, which implies that the spine curve is a cylindrical curve with the **k**-axis as the central axis. This has proved that the given circular surface is a $C'R$ workspace surface. □

Remark 3.7. Three initial conditions are required to determine the origin, but only two, C_1 and C_2, have been used in Eq. (3.41). Hence, the position of the spine curve is not totally decided. Since the choice of the origin has one DOF, we can choose an arbitrary point along the **k**-axis to fix the origin.

Remark 3.8. The D–H parameter of the $C'R$-linkage mechanism that generates the workspace surface can be obtained from Eq. (3.38) as

$$
\begin{aligned}
\delta_1 &= \operatorname{arccot} k_g \\
a_1 &= (\alpha_3 \cot k_g - \alpha_1 \sin k_g) \sin k_g
\end{aligned}
$$

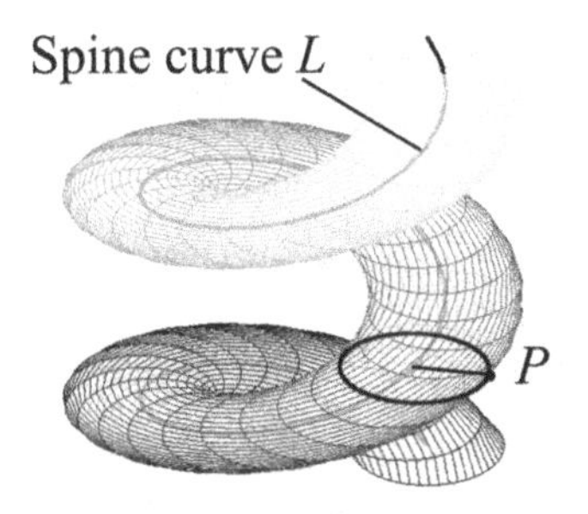

Fig. 3.24: An HR workspace surface

$$d_1 = \int_{u_0}^{u} (\alpha_3 \sin k_g + \alpha_1 \cos k_g) du$$

$$d_2 = \alpha_2$$

It is straightforward to have the following corollaries.

Corollary 3.1. *The necessary and sufficient condition for a circular surface to be an HR workspace surface is* (1) *the geodesic curvature k_g and $\alpha_i, i = 1, 2, 3$ are constants*; (2) $\alpha_1 - \alpha_3 k_g \neq 0$.

An HR workspace surface is shown in Fig. 3.24.

Corollary 3.2. *The necessary and sufficient condition for a circular surface to be an HR workspace surface is* (1) *the geodesic curvature k_g and $\alpha_i, i = 1, 2, 3$ are constants*; (2) $\alpha_1 - \alpha_3 k_g = 0$.

3.6 Conclusion

We have presented the examples of application of the adjoint approach to curves and surfaces. These examples include the involute and evolute curves and the roulettes of planar curves and the Bernard curves of spatial curves. We have introduced the fixed-point conditions for a planar curve, a spatial curve, and a surface curve. Finally we have demonstrated its application to a circular surface, which showcases a surface adjoint to a spatial curve.

Part II

Forward Kinematics of Rolling–Sliding Contact

Chapter 4

From Trajectories to Velocity: Forward Kinematics of Rigid Surfaces with Rolling Contact

In this chapter, we develop the forward kinematics of two rigid surfaces with rolling contact when the contact trajectory on each surface is present. We use the moving-frame method and the adjoint approach to obtain the instantaneous forward kinematics in terms of geometric invariants: rolling rate, normal curvature, geodesic curvature, and geodesic torsion. This approach is coordinate-invariant in the sense that the equations are invariant w.r.t. the Special Euclidean group in $\mathbb{R}^3$.

4.1 Problem Definition

We are to study the relative motion between two surfaces, so we assume one surface fixed and the other moving. The moving surface has three rotational DOFs about the contact point M: one spin motion about the normal vector and two rotational motions about two base vectors in the tangent plane, as shown in Fig. 4.1.

Let S denote the fixed surface, on which a frame $(\mathbf{O}, \mathbf{i}, \mathbf{j}, \mathbf{k})$ is fixed, and S' the moving surface, on which a frame $(\mathbf{O}', \mathbf{i}', \mathbf{j}', \mathbf{k}')$ is fixed.

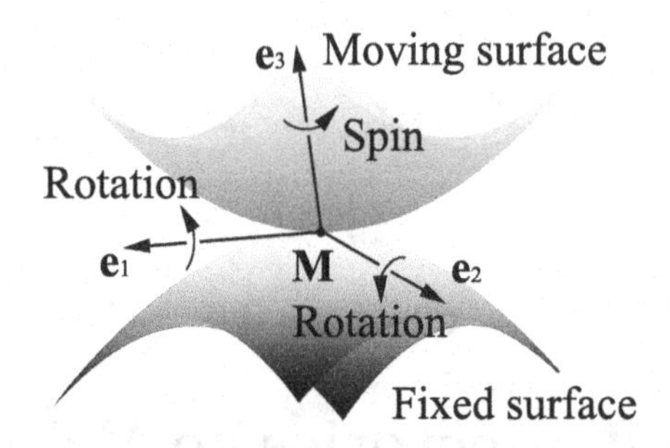

Fig. 4.1: The four DOFs of the moving surface

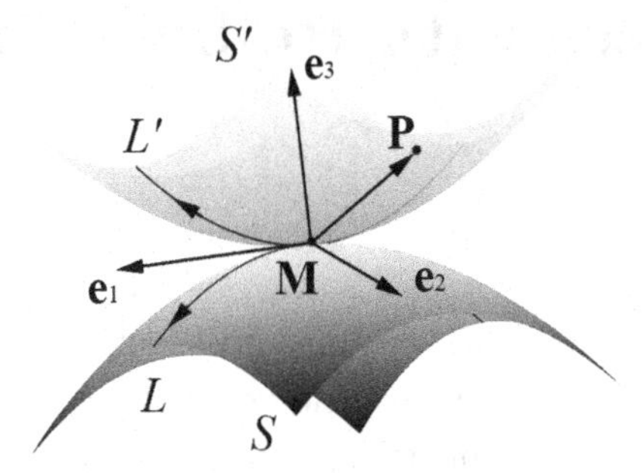

Fig. 4.2: The coinciding Darboux frames along the contact trajectories

Remark 4.1. Note the locations of these two frames are irrelevant due to the approach being coordinate-invariant, so we do not sketch these two frames in the figures in the book.

The rolling motion generates two contact trajectories—one on the fixed surface S and one on the moving surface S'—which are represented by L and L', respectively, as shown in Fig. 4.2.

We obtain the angular velocity of the moving surface S' when the two contact trajectory curves L and L' are present.

4.2 Geometric Velocity of an Adjoint Point on the Moving Surface

We use the following notation to specify the frame w.r.t. which a vector is referenced. A vector with a superscript prime is referenced w.r.t. the frame $(\mathbf{O}', \mathbf{i}', \mathbf{j}', \mathbf{k}')$; otherwise it is referenced w.r.t. the frame $(\mathbf{O}, \mathbf{i}, \mathbf{j}, \mathbf{k})$.

Let M denote the contact point on L and M' the contact point on L'. We attach the Darboux frame $(\mathbf{M}, \mathbf{e}_1, \mathbf{e}_2, \mathbf{e}_3)$ to the curve L, where $\mathbf{e}_1$ is the unit tangent vector to L and $\mathbf{e}_3$ is the normal to the surface S. We further attach the Darboux frame $(\mathbf{M}', \mathbf{e}_1', \mathbf{e}_2', \mathbf{e}_3')$ to the curve L', where $\mathbf{e}_1'$ is the tangent vector to L' and $\mathbf{e}_3'$ is the normal to the surface S'.

Two rigid surfaces with rolling contact entail three geometric constraints: The two contact points coincide. The normal vectors of the two surfaces at the contact points are collinear. The contact points have the same velocity.

Geometrically, we have the following constraints: the contact points M and M' coinciding, the tangent vectors $\mathbf{e}_1$ and $\mathbf{e}_1'$ collinear, and the normals $\mathbf{e}_3$ and $\mathbf{e}_3'$ collinear. We can hence make the two frames $(\mathbf{M}, \mathbf{e}_1, \mathbf{e}_2, \mathbf{e}_3)$ and $(\mathbf{M}', \mathbf{e}_1', \mathbf{e}_2', \mathbf{e}_3')$ coincide, as shown in Fig. 4.2.

4.2.1 *The Fixed-Point Conditions*

Let P denote an arbitrary fixed point on the moving surface S', and it serves as the adjoint point for the two contact trajectories L and L'.

Let (u_1', u_2', u_3') denote the coordinates of the point P referenced w.r.t. the frame $(\mathbf{M}', \mathbf{e}_1', \mathbf{e}_2', \mathbf{e}_3')$. The position vector $\mathbf{P}'$ referenced w.r.t. the frame $(\mathbf{O}', \mathbf{i}', \mathbf{j}', \mathbf{k}')$ can be expressed as

$$\mathbf{P}' = \mathbf{M}' + u_1'\mathbf{e}_1' + u_2'\mathbf{e}_2' + u_3'\mathbf{e}_3' \tag{4.1}$$

Let s' denote the arc length of L'. Differentiating both sides of Eq. (4.1) w.r.t. s' yields

$$\begin{aligned}\frac{d\mathbf{P}'}{ds'} &= \frac{d\mathbf{M}'}{ds'} + \frac{du_1'}{ds'}\mathbf{e}_1' + u_1'\frac{d\mathbf{e}_1'}{ds'} + \frac{du_2'}{ds'}\mathbf{e}_2' + u_2'\frac{d\mathbf{e}_2'}{ds'} + \frac{du_3'}{ds'}\mathbf{e}_3' + u_3'\frac{d\mathbf{e}_3'}{ds'} \\ &= \mathbf{e}_1' + \frac{du_1'}{ds'}\mathbf{e}_1' + u_1'(k_g'\mathbf{e}_2' + k_n'\mathbf{e}_3')\end{aligned}$$

$$
\begin{aligned}
&+\frac{du_2'}{ds'}\mathbf{e}_2' + u_2'(-k_g'\mathbf{e}_1' + \tau_g'\mathbf{e}_3') \\
&+\frac{du_3'}{ds'}\mathbf{e}_3' + u_3'(-k_n'\mathbf{e}_1' - \tau_g'\mathbf{e}_2') \\
&= \left(1 + \frac{du_1'}{ds'} - u_2'k_g' - u_3'k_n'\right)\mathbf{e}_1' \\
&+ \left(\frac{du_2'}{ds'} + u_1'k_g' - u_3'\tau_g'\right)\mathbf{e}_2' \\
&+ \left(\frac{du_3'}{ds'} + u_1'k_n' + u_2'\tau_g'\right)\mathbf{e}_3'
\end{aligned} \tag{4.2}
$$

where k_g' is the geodesic curvature, k_n' the normal curvature, τ_g' the geodesic torsion of the contact trajectory L', as in the Darboux formulas in Eq. (2.46).

Note that P is a fixed point on the surface S', so we can obtain the fixed-point conditions as in Eq. (3.28) in Chapter 3. Since $d\mathbf{P}'/ds' = 0$ w.r.t. the frame ($\mathbf{O}'$, $\mathbf{i}'$, $\mathbf{j}'$, $\mathbf{k}'$), the coefficients of $\mathbf{e}_i'$ on the right side of Eq. (4.2) are 0. The fixed-point conditions can be obtained as

$$
\begin{aligned}
\frac{du_1'}{ds'} &= u_2'k_g' + u_3'k_n' - 1 \\
\frac{du_2'}{ds'} &= -u_1'k_g' + u_3'\tau_g' \\
\frac{du_3'}{ds'} &= -u_1'k_n' - u_2'\tau_g'
\end{aligned} \tag{4.3}
$$

4.2.2 *The Geometric Velocity of the Adjoint Point*

Let (u_1, u_2, u_3) denote the coordinates of the point P referenced w.r.t. the frame ($\mathbf{M}$, $\mathbf{e}_1$, $\mathbf{e}_2$, $\mathbf{e}_3$). The position vector of the point P w.r.t. the frame ($\mathbf{O}$, $\mathbf{i}$, $\mathbf{j}$, $\mathbf{k}$) can be obtained as

$$
\mathbf{P} = \mathbf{M} + u_1\mathbf{e}_1 + u_2\mathbf{e}_2 + u_3\mathbf{e}_3 \tag{4.4}
$$

Let s represent arc length of L. Differentiating Eq. (4.4) w.r.t. s yields

$$\begin{aligned}\frac{d\mathbf{P}}{ds} &= \frac{d\mathbf{M}}{ds} + \frac{du_1}{ds}\mathbf{e}_1 + u_1\frac{d\mathbf{e}_1}{ds} + \frac{du_2}{ds}\mathbf{e}_2 + u_2\frac{d\mathbf{e}_2}{ds} + \frac{du_3}{ds}\mathbf{e}_3 + u_3\frac{d\mathbf{e}_3}{ds} \\ &= \mathbf{e}_1 + \frac{du_1}{ds}\mathbf{e}_1 + u_1(k_g\mathbf{e}_2 + k_n\mathbf{e}_3) \\ &\quad + \frac{du_2}{ds}\mathbf{e}_2 + u_2(-k_g\mathbf{e}_1 + \tau_g\mathbf{e}_3) \\ &\quad + \frac{du_3}{ds}\mathbf{e}_3 + u_3(-k_n\mathbf{e}_1 - \tau_g\mathbf{e}_2) \\ &= (1 + \frac{du_1}{ds} - u_2k_g - u_3k_n)\mathbf{e}_1 \\ &\quad + \left(\frac{du_2}{ds} + u_1k_g - u_3\tau_g\right)\mathbf{e}_2 \\ &\quad + \left(\frac{du_3}{ds} + u_1k_n + u_2\tau_g\right)\mathbf{e}_3 \end{aligned} \tag{4.5}$$

where k_g is the geodesic curvature, k_n the normal curvature, and τ_g the geodesic torsion of the contact trajectory L.

The rolling constraints entail that the contact points M and M' have the same velocity, thus the arc lengths of the two contact points covered in the same period of time are equal, i.e., $s = s'$. We have made the two Darboux frames ($\mathbf{M}$, $\mathbf{e}_1$, $\mathbf{e}_2$, $\mathbf{e}_3$) and ($\mathbf{M}'$, $\mathbf{e}'_1$, $\mathbf{e}'_2$, $\mathbf{e}'_3$) coincide. It follows that

$$u_1 = u'_1, \quad u_2 = u'_2, \quad u_3 = u'_3 \tag{4.6}$$

and consequently

$$\begin{aligned}\frac{du_1}{ds} &= \frac{du'_1}{ds'} \\ \frac{du_2}{ds} &= \frac{du'_2}{ds'} \\ \frac{du_3}{ds} &= \frac{du'_3}{ds'}\end{aligned} \tag{4.7}$$

We use s to denote both s and s' and u_i to denote both u_i and u_i', $i = 1, 2, 3$. Substituting Eqs. (4.3) and (4.7) into Eq. (4.5) yields the geometric velocity of the point P as

$$\frac{d\mathbf{P}}{ds} = (u_2 k_g^* + u_3 k_n^*)\mathbf{e}_1 + (-u_1 k_g^* + u_3 \tau_g^*)\mathbf{e}_2 + (-u_1 k_n^* - u_2 \tau_g^*)\mathbf{e}_3 \tag{4.8}$$

where

$$k_g^* = k_g' - k_g, \quad k_n^* = k_n' - k_n, \quad \tau_g^* = \tau_g' - \tau_g \tag{4.9}$$

We name k_g^* the induced geodesic curvature, k_n^* the induced normal curvature, and τ_g^* the induced geodesic torsion.

4.3 Angular Velocity of the Moving Surface

When time t is taken into consideration, the velocity of the point P on the moving object can be obtained from Eq. (4.8) as

$$\begin{aligned} \mathbf{v}_P &= \frac{d\mathbf{P}}{dt} = \frac{d\mathbf{P}}{ds}\frac{ds}{dt} \\ &= \sigma(u_2 k_g^* + u_3 k_n^*)\mathbf{e}_1 + \sigma(-u_1 k_g^* + u_3 \tau_g^*)\mathbf{e}_2 \\ &\quad + \sigma(-u_1 k_n^* - u_2 \tau_g^*)\mathbf{e}_3 \end{aligned} \tag{4.10}$$

where $\sigma = ds/dt$ represents the magnitude of the rolling velocity, i.e., the rolling rate.

Let $\boldsymbol{\omega}$ denote the angular velocity of the moving surface w.r.t. the frame ($\mathbf{O}$, $\mathbf{i}$, $\mathbf{j}$, $\mathbf{k}$). It can be decomposed into three components along the Darboux frame ($\mathbf{M}$, $\mathbf{e}_1$, $\mathbf{e}_2$, $\mathbf{e}_3$) as

$$\boldsymbol{\omega} = \omega_1 \mathbf{e}_1 + \omega_2 \mathbf{e}_2 + \omega_3 \mathbf{e}_3 \tag{4.11}$$

Let $\mathbf{r}_{MP}$ denote the position vector from the point M to point P w.r.t. the frame ($\mathbf{O}$, $\mathbf{i}$, $\mathbf{j}$, $\mathbf{k}$). It follows that

$$\mathbf{r}_{MP} = u_1 \mathbf{e}_1 + u_2 \mathbf{e}_2 + u_3 \mathbf{e}_3 \tag{4.12}$$

The velocity of the point P can also be obtained from Eqs. (4.11) and (4.12) as

$$\begin{aligned}\mathbf{v}_P &= \boldsymbol{\omega} \times \mathbf{r}_{MP} \\ &= (-u_2\omega_3 + u_3\omega_2)\mathbf{e}_1 + (u_1\omega_3 - u_3\omega_1)\mathbf{e}_2 \\ &\quad + (-u_1\omega_2 + u_2\omega_1)\mathbf{e}_3 \end{aligned} \tag{4.13}$$

The velocity of point P now has two expressions: Eqs. (4.10) and (4.13). They must be equal to each other. It follows that

$$\begin{aligned}\sigma(u_2k_g^* + u_3k_n^*) &= -u_2\omega_3 + u_3\omega_2 \\ \sigma(-u_1k_g^* + u_3\tau_g^*) &= u_1\omega_3 - u_3\omega_1 \\ \sigma(-u_1k_n^* - u_2\tau_g^*) &= -u_1\omega_2 + u_2\omega_1\end{aligned} \tag{4.14}$$

Solving Eq. (4.14) for ω_1, ω_2, and ω_3 yields

$$\omega_1 = -\sigma\tau_g^*, \quad \omega_2 = \sigma k_n^*, \quad \omega_3 = -\sigma k_g^* \tag{4.15}$$

The angular velocity $\boldsymbol{\omega}$ of the moving surface can thus be obtained from Eqs. (4.11) and (4.15) as

$$\boldsymbol{\omega} = \sigma(-\tau_g^*\mathbf{e}_1 + k_n^*\mathbf{e}_2 - k_g^*\mathbf{e}_3) \tag{4.16}$$

Remark 4.2. Angular acceleration of the moving surface can be straightforwardly obtained by differentiating Eq. (4.16).

Remark 4.3. The first two components in Eq. (3.16), $-\sigma\tau_g^*\mathbf{e}_1 + \sigma k_n^*\mathbf{e}_2$, give the pure-rolling velocity about an axis passing the contact point M in the tangent plane, the third component, $-\sigma k_g^*\mathbf{e}_3$ gives the spin velocity about the normal at the contact point M.

When the spin velocity is 0, it is called pure-rolling motion. It can be deduced from Eq. (4.16) that pure-rolling only occurs when $k_g = k_g'$. We thus have the following corollary:

Corollary 4.1. *Two surfaces undergo pure-rolling motion when the values of the geodesic curvature of the contact trajectories are equal.*

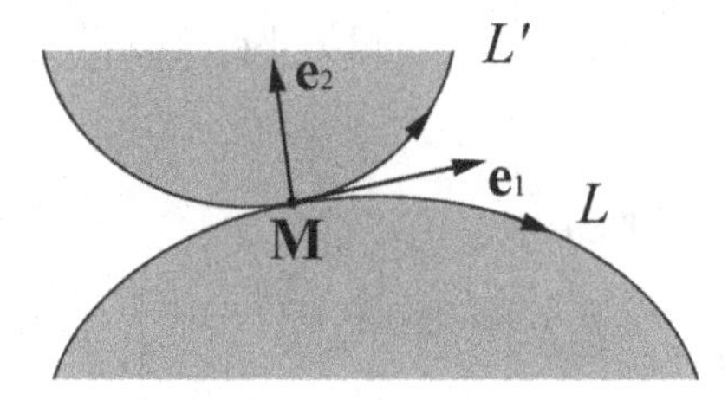

Fig. 4.3: The coinciding Darboux frames along the contact trajectories

4.4 Angular Velocity of Two Planar Laminas with Rolling Contact

In particular, this approach can also be applied to two planar rigid laminas maintaining rolling contact. The moving lamina has one rotational DOF about the normal vector of the laminas, which is a constant. The contact trajectories are two planar curves L and L', as shown in Fig. 4.3.

We have shown that both the normal curvature k_n and geodesic torsion τ_g are 0 and the geodesic curvature k_g is equal to the curvature k for a planar curve in Chapter 2. The angular velocity of the moving object can be obtained from Eq. (4.16) as

$$\omega = \sigma(k'_g - k_g)\mathbf{e}_3 = \sigma(k' - k)\mathbf{e}_3 \tag{4.17}$$

where k' denotes the curvature of L' and k that of L.

4.5 Reconciliation with a Classical Example

We show the reconciliation of Eq. (4.16) with the results of a classical example: a disk rolling on a plane.

Consider the classical example of an upright disk of radius R rolling on a plane. The contact trajectories are the circle L' of the disk and L in the plane, as shown in Fig. 4.4.

We set up the Darboux frames ($\mathbf{M}$, $\mathbf{e}_1$, $\mathbf{e}_2$, $\mathbf{e}_3$) along the curve L in the plane S, where $\mathbf{e}_1$ is the unit tangent vector and $\mathbf{e}_3$ is the normal to the plane and points upwards. We further set up the

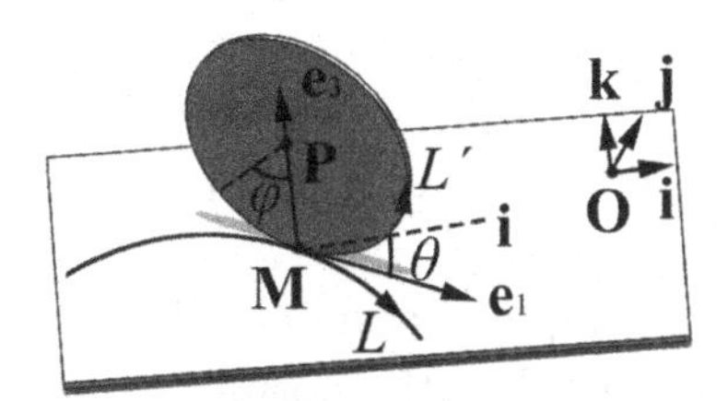

Fig. 4.4: An upright disk rolling on a plane

Darboux frame ($\mathbf{M}'$, $\mathbf{e}_1'$, $\mathbf{e}_2'$, $\mathbf{e}_3'$) along the curve L', where $\mathbf{e}_1'$ is the unit tangent vector and $\mathbf{e}_3'$ points to the circle center. These two Darboux frames are made to coincide.

Let s denote the arc length of L and k the curvature of L. Example 2.6 has shown the Darboux formulas of the planar curve L:

$$\frac{d}{ds}\begin{bmatrix}\mathbf{e}_1\\ \mathbf{e}_2\\ \mathbf{e}_3\end{bmatrix} = \begin{bmatrix}0 & k & 0\\ -k & 0 & 0\\ 0 & 0 & 0\end{bmatrix}\begin{bmatrix}\mathbf{e}_1\\ \mathbf{e}_2\\ \mathbf{e}_3\end{bmatrix} \tag{4.18}$$

The disk can be treated as a degenerated sphere, where a great circle is the contact trajectory curve and the normal $\mathbf{e}_3'$ points toward the sphere center. Example 2.7 has shown the Darboux formulas of a small circle. When the curve L' is a big circle, the geodesic curvature k_g becomes 0 because $u_0 = 0$, and the normal curvature k_n is $1/R$. Thus the Darboux formulas of the big circle L' are

$$\frac{d}{ds'}\begin{bmatrix}\mathbf{e}_1'\\ \mathbf{e}_2\\ \mathbf{e}_3'\end{bmatrix} = \begin{bmatrix}0 & 0 & \frac{1}{R}\\ 0 & 0 & 0\\ -\frac{1}{R} & 0 & 0\end{bmatrix}\begin{bmatrix}\mathbf{e}_1'\\ \mathbf{e}_2'\\ \mathbf{e}_3'\end{bmatrix} \tag{4.19}$$

The angular velocity of the disk can be obtained from Eq. (4.16) as

$$\boldsymbol{\omega} = \sigma\left(\frac{1}{R}\mathbf{e}_2 - k\mathbf{e}_3\right) \tag{4.20}$$

where σ is the rolling rate. The position vector $\mathbf{r}$ of the center point of the disk w.r.t. the frame ($\mathbf{M}$, $\mathbf{e}_1$, $\mathbf{e}_2$, $\mathbf{e}_3$) is $R\mathbf{e}_3$. The velocity of the center point is thus

$$\mathbf{v} = \boldsymbol{\omega} \times \mathbf{r} = \sigma \left(\frac{1}{R}\mathbf{e}_2 - k\mathbf{e}_3 \right) \times R\mathbf{e}_3 = \sigma \mathbf{e}_1 \tag{4.21}$$

In textbooks, for example [Rosenberg, 1977], a frame $(\mathbf{O}, \mathbf{i}, \mathbf{j}, \mathbf{k})$ is fixed to the plane, where $\mathbf{k}$ is the normal. Four general coordinates (x, y, θ, φ) are used to derive the kinematics, where θ is the angle between $\mathbf{i}$ and $\mathbf{e}_1$ and φ is the rolling angle of the disk. The angular velocity of the disk and the velocity of the center of the disk are obtained respectively as

$$\begin{aligned} \boldsymbol{\omega} &= \frac{d\varphi}{dt}\cos\theta\mathbf{i} + \frac{d\varphi}{dt}\sin\theta\mathbf{j} - \frac{d\theta}{dt}\mathbf{k} \\ \mathbf{v} &= R\frac{d\varphi}{dt}\cos\theta\mathbf{i} + R\frac{d\varphi}{dt}\sin\theta\mathbf{j} \end{aligned} \tag{4.22}$$

The following relationships exist:

$$\begin{aligned} \mathbf{e}_1 &= \cos\theta\mathbf{i} + \sin\theta\mathbf{j} \\ \mathbf{e}_2 &= -\sin\theta\mathbf{i} + \cos\theta\mathbf{j} \\ \mathbf{e}_3 &= \mathbf{k} \\ \sigma &= \frac{ds}{dt} = R\frac{d\varphi}{dt} \\ k &= \frac{d\theta}{ds} \end{aligned} \tag{4.23}$$

The angular velocity of the disk and translational velocity of the center point in these two formulations can be verified to be the same after Eq. (4.23) is substituted into Eqs. (4.20) and (4.21).

4.6 Examples of Application

We use some examples to show how to obtain the kinematics of the moving surface when the contact trajectories are present.

Example 4.1. Consider a sphere of unit radius rolling on a plane. The contact trajectory curve L in the plane has curvature k and the

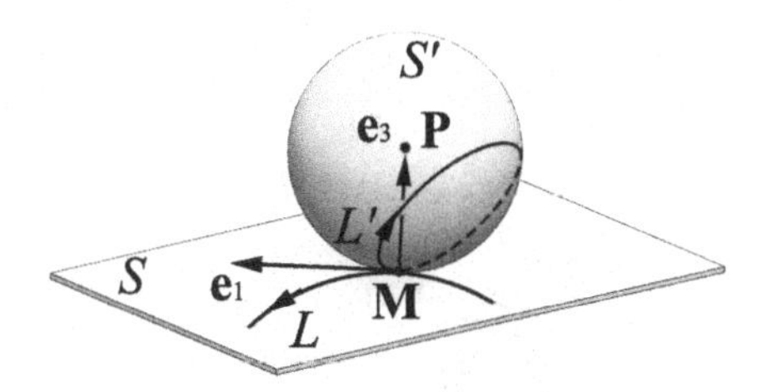

Fig. 4.5: A sphere rolling on a plane

contact trajectory curve L' on the sphere is a small circle, as shown in Fig. 4.5.

Solution:
We set up the Darboux frame $(\mathbf{M}, \mathbf{e}_1, \mathbf{e}_2, \mathbf{e}_3)$, where $\mathbf{e}_1$ is the common tangent vector to L and L' and $\mathbf{e}_3$ is the common normal and points to the sphere center P.

The curvatures of L are given in Example 2.6:

$$k_g = k, \quad k_n = 0, \quad \tau_g = 0$$

The curvatures of L' are given in Example 2.7:

$$k'_g = \tan u_0, \quad k'_n = 1, \quad \tau'_g = 0$$

The induced curvatures are

$$k^*_g = \tan u_0 - k, \quad k^*_n = 1, \quad \tau^*_g = 0$$

The angular velocity of the sphere can be obtained from Eq. (4.16) as

$$\boldsymbol{\omega} = \sigma\,(\mathbf{e}_2 - (\tan u_0 - k)\mathbf{e}_3)$$

The position vector of the center point P w.r.t. the Darboux frame is $\mathbf{r} = \mathbf{e}_3$. The velocity of the center point is thus

$$\mathbf{v} = \boldsymbol{\omega} \times \mathbf{r} = \sigma \mathbf{e}_1$$ □

Example 4.2. Consider a paraboloid rolling on a cylinder. Suppose the radius of the cylinder is R and the contact trajectory L is the

Fig. 4.6: A parabolic rolling on a cylinder

intersecting curve of the cylinder and a sphere of radius $2R$. The paraboloid is generated by rotating a parabola $z = x^2/2$ about the z-axis and the contact trajectory curve L' is one of the meridians, as shown in Fig. 4.6.

Solution:

We set up the Darboux frame ($\mathbf{M}$, $\mathbf{e}_1$, $\mathbf{e}_2$, $\mathbf{e}_3$) such that $\mathbf{e}_1$ is the common tangent vector to L and L' and $\mathbf{e}_3$ is the common normal pointing outward of the cylinder.

The cylinder can be parameterized as

$$\mathbf{x}(u, v) = R\left[\cos u,\ \sin u,\ v\right]$$

The contact trajectory curve L can be parameterized as

$$\mathbf{M}(u) = R\left[\cos u,\ \sin u,\ 2\sin\tfrac{u}{2}\right]$$

The unit tangent vector $\mathbf{e}_1$ can be obtained as

$$\mathbf{e}_1 = \frac{d\mathbf{M}}{ds} = \frac{d\mathbf{M}/du}{|d\mathbf{M}/du|} = \frac{\left[-\sin u,\ \cos u,\ \cos\tfrac{u}{2}\right]}{\sqrt{1+\cos^2\tfrac{u}{2}}}$$

Pointing outward, the normal $\mathbf{e}_3$ can be obtained as

$$\mathbf{e}_3 = \left[\cos u,\ \sin u,\ 0\right]$$

The geodesic curvature k_g, normal curvature k_n, and geodesic torsion τ_g of the Darboux frame can be obtained from Eq. (2.46) as

$$\begin{aligned} k_g &= \frac{d\mathbf{e}_1}{ds} \cdot \mathbf{e}_2 = \frac{-\sin\frac{u}{2}}{2R(1+\cos^2\frac{u}{2})^{\frac{3}{2}}} \\ k_n &= \frac{d\mathbf{e}_1}{ds} \cdot \mathbf{e}_3 = \frac{-1}{R(1+\cos^2\frac{u}{2})^{\frac{3}{2}}} \\ \tau_g &= \frac{d\mathbf{e}_2}{ds} \cdot \mathbf{e}_3 = \frac{\cos\frac{u}{2}}{R(1+\cos^2\frac{u}{2})^{\frac{3}{2}}} \end{aligned} \tag{4.24}$$

The paraboloid can be parameterized as

$$\mathbf{x}'(\alpha, \beta) = \left[\alpha,\ \beta,\ \tfrac{1}{2}(\alpha^2+\beta^2)\right]$$

The contact trajectory curve L' can be parameterized as

$$\mathbf{M}'(\alpha) = \left[\alpha,\ \beta_0,\ \tfrac{1}{2}(\alpha^2+\beta_0^2)\right]$$

where β_0 is a constant. The unit tangent vector $\mathbf{e}_1'$ can be obtained as

$$\mathbf{e}_1' = \frac{d\mathbf{M}'}{ds'} = \frac{d\mathbf{M}'/d\alpha}{|d\mathbf{M}'/d\alpha|} = \frac{[1,\ 0,\ \alpha]}{\sqrt{\alpha^2+1}}$$

Pointing inward, the normal $\mathbf{e}_3'$ can be obtained as

$$\mathbf{e}_3' = \frac{[\alpha,\ \beta_0,\ -1]}{\sqrt{\alpha^2+\beta_0^2+1}}$$

The geodesic curvature k_g', normal curvature k_n', and geodesic torsion τ_g' of the Darboux frame can be obtained from Eq. (2.33) as

$$\begin{aligned} k_g' &= \frac{d\mathbf{e}_1'}{ds} \cdot \mathbf{e}_2' = \frac{-\beta_0}{(\alpha^2+1)^{\frac{3}{2}}\sqrt{\alpha^2+\beta_0^2+1}} \\ k_n' &= \frac{d\mathbf{e}_1'}{ds} \cdot \mathbf{e}_3' = \frac{-1}{(\alpha^2+1)\sqrt{\alpha^2+\beta_0^2+1}} \\ \tau_g' &= \frac{d\mathbf{e}_2'}{ds} \cdot \mathbf{e}_3' = \frac{-\alpha\beta_0}{(\alpha^2+1)\sqrt{\alpha^2+\beta_0^2+1}} \end{aligned} \tag{4.25}$$

The angular velocity of the paraboloid can be obtained by substituting Eqs. (4.24) and (4.25) into Eq. (4.16), and its expression will be skipped here. □

4.7 Application to Qualitative Trajectory Analysis

Qualitative information about necessary contact trajectories can be deduced if some constraints, such as motion or shapes of surfaces, are taken into consideration.

Recall that the properties of a surface can be partially revealed through its curves. Some curves have special continuous geometric characteristics. For instance, a line of curvature is a regular connected curve such that the geodesic torsion τ_g vanishes. A geodesic is a regular connected curve such that the geodesic curvature k_g vanishes. These properties can be used for qualitative trajectory analysis.

4.7.1 *Trajectory Analysis of Pure-Rolling*

An object undergoes pure-rolling when the angular velocity in the direction of the contact normal vanishes. Corollary 4.1 reveals that the values of the geodesic curvatures of two contact trajectories have to be equal for the moving object to undergo pure-rolling.

Geodesics serve as contact trajectories naturally generate pure-rolling motions since $k_g = k_g' = 0$, leading to 0 angular velocity in the direction of normal. Further, the distance between two configurations is the shortest if the surfaces are complete, which is often true for objects under consideration in this book.

It is often true that the shapes of the objects are known in advance and a certain movement of the moving object is preferred. In this case, qualitative information about the contact curve on the fixed object can be deduced.

A sphere with radius R undergoes pure-rolling on a plane along a small circle L' on the sphere and a curve L in the plane, as shown in Fig. 4.5. Example 3.1 gives the angular velocity of the sphere as

$$\boldsymbol{\omega} = \sigma\left(\mathbf{e}_2 - (\tan u_0 - k)\mathbf{e}_3\right)$$

If the sphere undergoes pure-rolling, the $\mathbf{e}_3$ component of the angular velocity must be 0. It follows that

$$k = \tan u_0$$

This means the curve L has to be a circle with radius $1/\tan u_0$, as shown in Fig. 4.7.

4.7.2 *Trajectory Analysis Based on Shape Characteristics*

Even if an object undergoes the more general spin–rolling, qualitative information about the contact trajectories can be deduced. Such information is valuable when trajectory analysis is performed.

A disk of unit radius rolls on a surface S, maintaining upright at all time, as shown in Fig. 4.8. The constraint of remaining upright means that the normal vector of the surface always points to the center of the disk. It can also be imagined as a ball rolling on a smooth surface along one great circle.

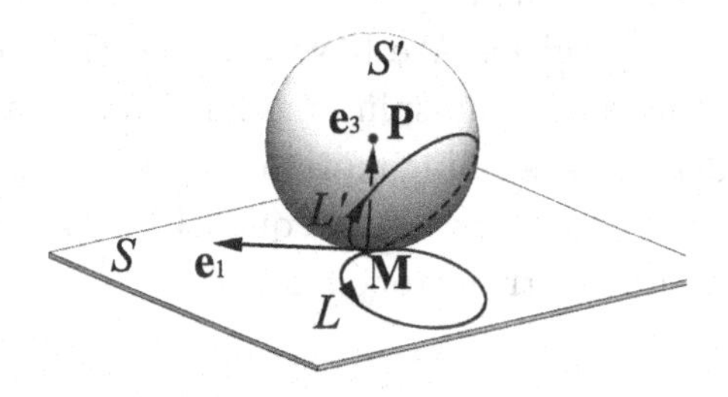

Fig. 4.7: The circular contact trajectory in the plane

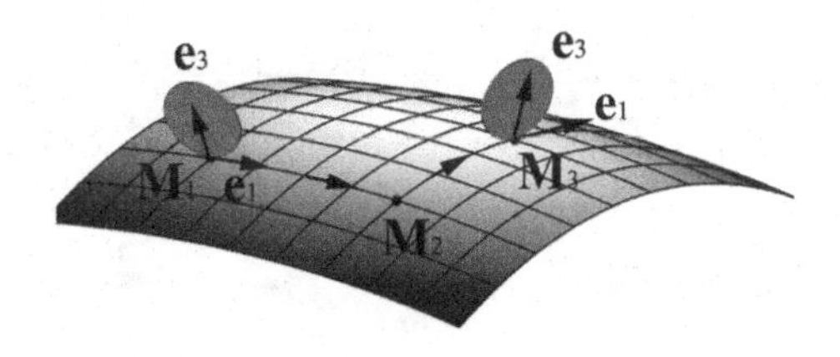

Fig. 4.8: The path for a disk on a surface

Let k_g represent the geodesic curvature, k_n the normal curvature, and τ_g the geodesic torsion of a contact trajectory on the surface. The angular velocity of the disk can be obtained from Eq. (4.16) as

$$\boldsymbol{\omega} = \sigma\left(\tau_g \mathbf{e}_1 + (1 - k_n)\mathbf{e}_2 + k_g \mathbf{e}_3\right)$$

The disk maintaining upright implies that the $\mathbf{e}_1$ component of the angular velocity $\boldsymbol{\omega}$ is 0. It follows that the contact trajectory on the surface has to be one of the lines of curvature.

The black curves in Fig. 4.8 mark the lines of curvature of the surface. If the disk is to roll uprightly from position $\mathbf{M}_1$ to $\mathbf{M}_3$, it can follow the segment from $\mathbf{M}_1$ to $\mathbf{M}_2$ along one line of curvature and then along the segment from $\mathbf{M}_2$ to $\mathbf{M}_3$ along other line of curvature. There would be infinitely many paths the disk can follow as long as the paths are lines of curvature.

4.8 Conclusion

We have applied the fixed-point conditions of a point adjoint to a surface to obtaining the angular velocity of the moving surface when two contact curves are given. We have clarified the contributions of the geodesic curvatures to the spin angular velocity and found that two surfaces undergo pure-rolling motion only when the geodesic curvatures of the two contact curves are equal. Finally, we have applied the results to qualitative trajectory analysis.

Chapter 5

From Trajectories to Velocity: Forward Kinematics of Rigid Surfaces with Rolling–Sliding Contact

In this chapter, we extend the approach presented in Chapter 4 to rolling–sliding contact: the moving surface rolls and slides simultaneously on the fixed surface. We derive both the angular velocity and translational velocity of the moving surface when the two contact trajectories are present.

5.1 Problem Definition

When two rigid surfaces maintain rolling–sliding contact, the moving surface has five DOFs: three rotational DOFs about the contact point M and two translational DOFs of the contact point in the tangent plane, as shown in Fig. 5.1.

Let S denote the fixed surface and S' the moving object. The rolling–sliding motion generates two contact trajectories: L on the fixed surface S, which is generated by both the rolling and sliding motions, and L' on the moving surface S', which is generated solely by the rolling motion.

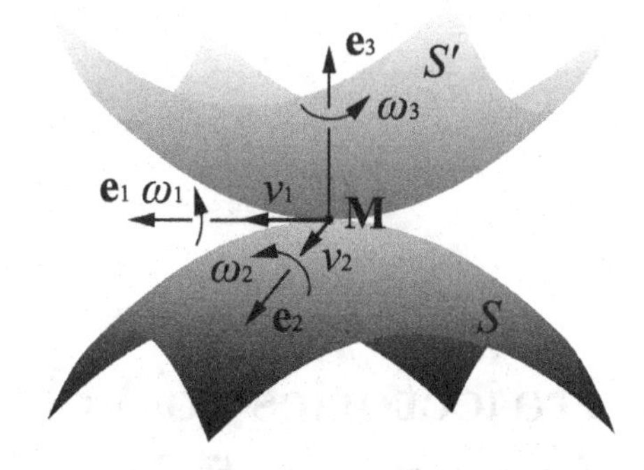

Fig. 5.1: The five DOFs of the moving object

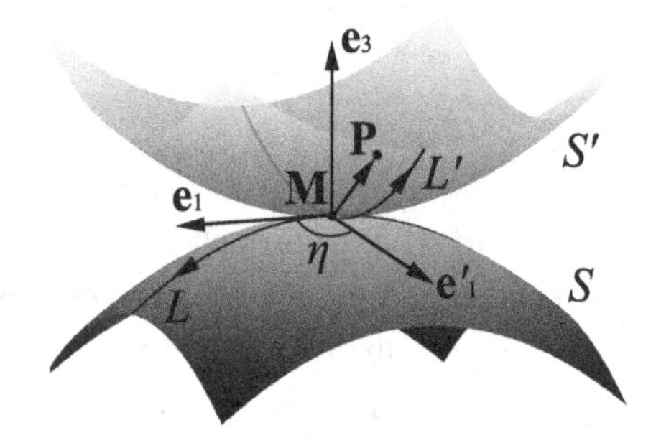

Fig. 5.2: The two Darboux frames along the contact trajectories

Let M denote the contact point on L and M' the contact point on L'. We attach the Darboux frame ($\mathbf{M}$, $\mathbf{e}_1$, $\mathbf{e}_2$, $\mathbf{e}_3$) to the curve L, where $\mathbf{e}_1$ is the unit tangent vector to L and $\mathbf{e}_3$ is the normal to the surface S. We further attach the Darboux frame ($\mathbf{M}'$, $\mathbf{e}_1'$, $\mathbf{e}_2'$, $\mathbf{e}_3'$) to the curve L', where $\mathbf{e}_1'$ is the tangent vector to L' and $\mathbf{e}_3'$ is the normal to the surface S'.

Two rigid surfaces with rolling–sliding contact entail two geometric constraints: First, the two contact points coincide. Second, the normal vectors of the two surfaces at the contact point are collinear. We hence have the contact points M and M' coinciding and $\mathbf{e}_3$ and $\mathbf{e}_3'$ collinear, as shown in Fig. 5.2.

We obtain the velocity of the moving surface when the two contact trajectories L and L' are present.

5.2 The Velocity of an Arbitrary Point on the Moving Surface

Let S' represent the arc length of L' and P denote an arbitrary point on S'. The position vector of the point P can be referenced w.r.t. the frame ($\mathbf{O}'$, $\mathbf{i}'$, $\mathbf{j}'$, $\mathbf{k}'$), which is fixed on S', as

$$\mathbf{P}' = \mathbf{M}' + u_1'\mathbf{e}_1' + u_2'\mathbf{e}_2' + u_3'\mathbf{e}_3' \tag{5.1}$$

Differentiating Eq. (5.1) w.r.t. S' yields

$$\begin{aligned}\frac{d\mathbf{P}'}{ds'} &= \left(1 + \tfrac{du_1'}{ds'} - u_2'k_g' - u_3'k_n'\right)\mathbf{e}_1' \\ &\quad + \left(\tfrac{du_2'}{ds'} + u_1'k_g' - u_3'\tau_g'\right)\mathbf{e}_2' \\ &\quad + \left(\tfrac{du_3'}{ds'} + u_1'k_n' + u_2'\tau_g'\right)\mathbf{e}_3'\end{aligned} \tag{5.2}$$

where k_g' represents the geodesic curvature, k_n' the normal curvature, and τ_g' the geodesic torsion of L'.

The point P being fixed on S' indicates $d\mathbf{P}'/ds' = 0$. It follows that

$$\begin{aligned}\frac{du_1'}{ds'} &= u_2'k_g' + u_3'k_n' - 1 \\ \frac{du_2'}{ds'} &= -u_1'k_g' + u_3'\tau_g' \\ \frac{du_3'}{ds'} &= -u_1'k_n' - u_2'\tau_g'\end{aligned} \tag{5.3}$$

Equation (5.3) can be written in matrix form as

$$\frac{d\mathbf{u}'}{ds'} = \boldsymbol{\Omega}'\mathbf{u}' - \mathbf{b} \tag{5.4}$$

where

$$\mathbf{u}' = \begin{bmatrix} u_1' \\ u_2' \\ u_3' \end{bmatrix}, \quad \boldsymbol{\Omega}' = \begin{bmatrix} 0 & k_g' & k_n' \\ -k_g' & 0 & \tau_g' \\ -k_n' & -\tau_g' & 0 \end{bmatrix}, \quad \mathbf{b} = \begin{bmatrix} 1 \\ 0 \\ 0 \end{bmatrix}$$

On the other hand, the point P can be referenced w.r.t. the frame $(\mathbf{O}, \mathbf{i}, \mathbf{j}, \mathbf{k})$, which is fixed on S, as

$$\mathbf{P} = \mathbf{M} + u_1\mathbf{e}_1 + u_2\mathbf{e}_2 + u_3\mathbf{e}_3 \tag{5.5}$$

Differentiating Eq. (5.5) w.r.t. s yields

$$\begin{aligned}\frac{d\mathbf{P}}{ds} = &\left(1 + \tfrac{du_1}{ds} - u_2k_g - u_3k_n\right)\mathbf{e}_1\\ &+ \left(\tfrac{du_2}{ds} + u_1k_g - u_3\tau_g\right)\mathbf{e}_2\\ &+ \left(\tfrac{du_3}{ds} + u_1k_n + u_2\tau_g\right)\mathbf{e}_3\end{aligned} \tag{5.6}$$

where k_g represents the geodesic curvature, k_n the normal curvature, and τ_g the geodesic torsion of L.

The frames $(\mathbf{M}', \mathbf{e}_1', \mathbf{e}_2', \mathbf{e}_3')$ and $(\mathbf{M}, \mathbf{e}_1, \mathbf{e}_2, \mathbf{e}_3)$ can be made to coincide by rotating the frame $(\mathbf{M}, \mathbf{e}_1, \mathbf{e}_2, \mathbf{e}_3)$ by an angle η about the $\mathbf{e}_3$-axis. It follows that

$$\begin{aligned}\mathbf{e}_1' &= \cos\eta\mathbf{e}_1 + \sin\eta\mathbf{e}_2\\ \mathbf{e}_2' &= -\sin\eta\mathbf{e}_1 + \cos\eta\mathbf{e}_2\\ \mathbf{e}_3' &= \mathbf{e}_3\end{aligned} \tag{5.7}$$

The position vector from the contact point M to the point P can be referenced w.r.t. two frames: $(\mathbf{M}, \mathbf{e}_1, \mathbf{e}_2, \mathbf{e}_3)$ and $(\mathbf{M}', \mathbf{e}_1', \mathbf{e}_2', \mathbf{e}_3')$. The two expressions must be equal. It follows that

$$u_1\mathbf{e}_1 + u_2\mathbf{e}_2 + u_3\mathbf{e}_3 = u_1'\mathbf{e}_1' + u_2'\mathbf{e}_2' + u_3'\mathbf{e}_3' \tag{5.8}$$

The relationship between u_i and u_i' can be obtained by substituting Eq. (5.7) into Eq. (5.8) as

$$\begin{aligned}u_1 &= u_1'\cos\eta - u_2'\sin\eta\\ u_2 &= u_1'\sin\eta + u_2'\cos\eta\\ u_3 &= u_3'\end{aligned} \tag{5.9}$$

Equation (5.9) can be written in matrix form as

$$\mathbf{u} = \mathbf{R}\mathbf{u}' \tag{5.10}$$

where

$$\mathbf{u} = \begin{bmatrix} u_1 \\ u_2 \\ u_3 \end{bmatrix}, \quad \mathbf{R} = \begin{bmatrix} \cos\eta & -\sin\eta & 0 \\ \sin\eta & \cos\eta & 0 \\ 0 & 0 & 1 \end{bmatrix}, \quad \mathbf{u}' = \begin{bmatrix} u_1' \\ u_2' \\ u_3' \end{bmatrix}$$

The contact trajectory curve L on S is generated by the rolling–sliding motion, and the contact trajectory curve L' on S' is generated by the rolling motion alone. In general, the arc length s of L is not equal to the arc length S' of L', so the rolling–sliding rate $\sigma = ds/dt$ is not equal to the rolling rate $\sigma' = ds'/dt$. We use λ to represent the ratio of σ' to σ, i.e., $\lambda = \sigma'/\sigma = ds'/ds$.

Differentiating Eq. (5.10) w.r.t. s yields

$$\begin{aligned} \frac{d\mathbf{u}}{ds} &= \frac{d(\mathbf{R}\mathbf{u}')}{ds} = \frac{ds'}{ds}\frac{d(\mathbf{R}\mathbf{u}')}{ds'} \\ &= \lambda\left(\frac{d\mathbf{R}}{ds'}\mathbf{u}' + \mathbf{R}\frac{d\mathbf{u}'}{ds'}\right) \\ &= \lambda\left(\frac{d\mathbf{R}}{ds'}(\mathbf{R}^{-1}\mathbf{u}) + \mathbf{R}\frac{d\mathbf{u}'}{ds'}\right) \end{aligned} \tag{5.11}$$

Substituting Eq. (5.4) into Eq. (5.11) yields

$$\begin{aligned} \frac{d\mathbf{u}}{ds} &= \lambda\left(\frac{d\mathbf{R}}{ds'}(\mathbf{R}^{-1}\mathbf{u}) + \mathbf{R}(\boldsymbol{\Omega}'\mathbf{u}' - \mathbf{b})\right) \\ &= \lambda\left(\frac{d\mathbf{R}}{ds'}(\mathbf{R}^{-1}\mathbf{u}) + \mathbf{R}(\boldsymbol{\Omega}'(\mathbf{R}^{-1}\mathbf{u}) - \mathbf{b})\right) \end{aligned} \tag{5.12}$$

Equation (5.12) gives the geometric velocity of the point P referenced w.r.t. the frame ($\mathbf{M}$, $\mathbf{e}_1$, $\mathbf{e}_2$, $\mathbf{e}_3$). The three components of the velocity can be obtained as

$$\begin{aligned} \frac{du_1}{ds} &= \lambda\left(\left(k_g' - \frac{d\eta}{ds'}\right)u_2 + (k_n'\cos\eta - \tau_g'\sin\eta)u_3 - \cos\eta\right) \\ \frac{du_2}{ds} &= \lambda\left(\left(-k_g' + \frac{d\eta}{ds'}\right)u_1 + (\tau_g'\cos\eta + k_n'\sin\eta)u_3 - \sin\eta\right) \\ \frac{du_3}{ds} &= \lambda\left((\tau_g'\sin\eta - k_n'\cos\eta)u_1 - (\tau_g'\cos\eta + k_n'\sin\eta)\,u_2\right) \end{aligned} \tag{5.13}$$

Substituting Eq. (5.13) into Eq. (5.6) yields the geometric velocity of the point P as

$$\frac{d\mathbf{P}}{ds} = A_1\mathbf{e}_1 + A_2\mathbf{e}_2 + A_3\mathbf{e}_3 \tag{5.14}$$

where

$$\begin{aligned}
A_1 &= 1 - \lambda\cos\eta - \left(k_g + \lambda\left(\frac{d\eta}{ds'} - k'_g\right)\right)u_2 \\
&\quad - (k_n - \lambda(k'_n\cos\eta - \tau'_g\sin\eta))u_3 \\
A_2 &= -\lambda\sin\eta + \left(k_g + \lambda\left(\frac{d\eta}{ds'} - k'_g\right)\right)u_1 \\
&\quad - (\tau_g - \lambda(k'_n\sin\eta + \tau'_g\cos\eta))u_3 \\
A_3 &= (k_n - \lambda(k'_n\cos\eta - \tau'_g\sin\eta))u_1 \\
&\quad + (\tau_g - \lambda(k'_n\sin\eta + \tau'_g\cos\eta))u_2
\end{aligned}$$

If time t is taken into consideration, the velocity of the point P referenced to the frame $(\mathbf{O}, \mathbf{i}, \mathbf{j}, \mathbf{k})$, which is fixed on S, can be obtained from Eq. (5.14) as

$$\frac{d\mathbf{P}}{dt} = \frac{ds}{dt}\frac{d\mathbf{P}}{ds} = \sigma(A_1\mathbf{e}_1 + A_2\mathbf{e}_2 + A_3\mathbf{e}_3) \tag{5.15}$$

5.3 The Velocity of the Moving Surface

The moving surface S' has two translational DOFs and three rotational DOFs. We decompose the translational velocity of the contact point M along the directions $\mathbf{e}_1$ and $\mathbf{e}_2$ and the rotational velocity along the directions $\mathbf{e}_1$, $\mathbf{e}_2$, and $\mathbf{e}_3$. The velocity of the moving surface can be represented as

$$\begin{aligned}
\mathbf{v} &= v_1\mathbf{e}_1 + v_2\mathbf{e}_2 \\
\boldsymbol{\omega} &= \omega_1\mathbf{e}_1 + \omega_2\mathbf{e}_2 + \omega_3\mathbf{e}_3
\end{aligned} \tag{5.16}$$

Let $\mathbf{r}_{MP}$ represent the position vector from the point M to the point P. The velocity of the point P can be obtained as

$$\begin{aligned}\frac{d\mathbf{P}}{dt} &= \mathbf{v} + \boldsymbol{\omega} \times \mathbf{r}_{MP} \\ &= \mathbf{v} + \boldsymbol{\omega} \times (u_1\mathbf{e}_1 + u_2\mathbf{e}_2 + u_3\mathbf{e}_3) \\ &= v_1\mathbf{e}_1 + v_2\mathbf{e}_2 + (-u_2\omega_3 + u_3\omega_2)\mathbf{e}_1 \\ &\quad + (-u_1\omega_3 - u_3\omega_1)\mathbf{e}_2 + (-u_1\omega_2 + u_2\omega_1)\mathbf{e}_3 \end{aligned} \tag{5.17}$$

Equating Eqs. (5.15) and (5.17) yields the following three equations:

$$\begin{aligned} &\sigma\left(1 - \lambda\cos\eta - \left(k_g + \lambda\left(\frac{d\eta}{ds'} - k_g'\right)\right)u_2\right. \\ &\qquad \left. -(k_n - \lambda(k_n'\cos\eta - \tau_g'\sin\eta))u_3\right) \\ &\quad = v_1 + (-u_2\omega_3 + u_3\omega_2) \\ &\sigma\left(-\lambda\sin\eta + \left(k_g + \lambda\left(\frac{d\eta}{ds'} - k_g'\right)\right)u_1\right. \\ &\qquad \left. -(\tau_g - \lambda(k_n'\sin\eta + \tau_g'\cos\eta))u_3\right) \\ &\quad = v_2 + (-u_1\omega_3 - u_3\omega_1) \\ &\sigma((k_n - \lambda(k_n'\cos\eta - \tau_g'\sin\eta))u_1 \\ &\qquad + (\tau_g - \lambda(k_n'\sin\eta + \tau_g'\cos\eta))u_2) \\ &\quad = -u_1\omega_2 + u_2\omega_1 \end{aligned} \tag{5.18}$$

Since P is an arbitrary point, u_i can take arbitrary values. Solving Eq. (5.18) for v_i and ω_i yields

$$\begin{aligned} v_1 &= \sigma(1 - \lambda\cos\eta) = \sigma - \sigma'\cos\eta \\ v_2 &= \sigma(-\lambda\sin\eta) = -\sigma'\sin\eta \\ \omega_1 &= \sigma\tau_g - \sigma'(k_n'\sin\eta + \tau_g'\cos\eta) \\ \omega_2 &= -\sigma k_n + \sigma'(k_n'\cos\eta - \tau_g'\sin\eta) \\ \omega_3 &= \sigma k_g - \sigma' k_g' + \frac{d\eta}{dt} \end{aligned}$$

where

$$\sigma' = \sigma\lambda, \quad \frac{d\eta}{dt} = \sigma' \frac{d\eta}{ds'}$$

Hence, the velocity of the moving surface S' is

$$\begin{aligned} \mathbf{v} &= (\sigma - \sigma' \cos\eta)\mathbf{e}_1 - \sigma' \sin\eta \mathbf{e}_2 \\ \boldsymbol{\omega} &= (\sigma\tau_g - \sigma'(k'_n \sin\eta + \tau'_g \cos\eta))\mathbf{e}_1 \\ &\quad + (-\sigma k_n + \sigma'(k'_N \cos\eta - \tau' u_g \sin\eta))\mathbf{e}_2 \\ &\quad + \left(\sigma k_g - \sigma' k'_g + \frac{d\eta}{dt}\right)\mathbf{e}_3 \end{aligned} \tag{5.19}$$

Equation (5.19) gives the velocity of the moving surface S', where $\mathbf{v}$ represents the sliding velocity of the contact point M, and $\boldsymbol{\omega}$ represents the rotational velocity about the contact point M.

5.4 Rolling–Slipping Contact as a Special Case

Rolling–slipping contact occurs when the sliding motion is in the same direction of rolling. In this case, the angle η and $d\eta/dt$ become 0 in Eq. (5.19). The velocity of the moving surface becomes:

$$\begin{aligned} \mathbf{v} &= (\sigma - \sigma')\mathbf{e}_1 \\ \boldsymbol{\omega} &= (\sigma\tau_g - \sigma'\tau'_g)\mathbf{e}_1 + (-\sigma k_n + \sigma' k'_n)\mathbf{e}_2 + (\sigma k_g - \sigma' k'_g)\mathbf{e}_3 \end{aligned} \tag{5.20}$$

If the slipping motion is negligible, the rolling–slipping contact becomes rolling contact, then $\sigma = \sigma'$. The velocity of the moving surface is

$$\begin{aligned} \mathbf{v} &= 0 \\ \boldsymbol{\omega} &= \sigma(-(\tau'_g - \tau_g)\mathbf{e}_1 + (k'_n - k_n)\mathbf{e}_2 - (k'_g - k_g)\mathbf{e}_3) \end{aligned} \tag{5.21}$$

Equation (5.21) is the same as Eq. (4.16), which is as expected.

5.5 The Influence of the Instantaneous Imprint Curve on the Angular Velocity

At any instant, the rolling motion produces an instantaneous imprint curve L_m on the fixed surface. Since the angular velocity of the moving surface is subject to the geometric properties of the imprint curve L_m, it is theoretically interesting to obtain its geometric properties to cast insight into the causes of angular velocity of the moving surface.

We set up the Darboux frame ($\mathbf{M}$, $\mathbf{e}_{1m}$, $\mathbf{e}_{2m}$, $\mathbf{e}_{3m}$) on the imprint curve L_m, and this frame coincides with the frame ($\mathbf{M}'$, $\mathbf{e}'_1$, $\mathbf{e}'_2$, $\mathbf{e}'_3$) due to the constraints of the rolling contact. It follows that $\mathbf{e}_{1m}$ and $\mathbf{e}_1$ make an angle η and $\mathbf{e}_{3m} = \mathbf{e}_3$, as shown in Fig. 5.3.

The frame ($\mathbf{M}$, $\mathbf{e}_{1m}$, $\mathbf{e}_{2m}$, $\mathbf{e}_{3m}$) can be obtained by rotating the frame ($\mathbf{M}$, $\mathbf{e}_1$, $\mathbf{e}_2$, $\mathbf{e}_3$) about $\mathbf{e}_3$ by the angle η. It follows that

$$\begin{aligned}
\mathbf{e}_{1m} &= \cos\eta\mathbf{e}_1 + \sin\eta\mathbf{e}_2 \\
\mathbf{e}_{2m} &= -\sin\eta\mathbf{e}_1 + \cos\eta\mathbf{e}_2 \\
\mathbf{e}_{3m} &= \mathbf{e}_3
\end{aligned} \tag{5.22}$$

Equation (5.22) can be written in matrix form as

$$\mathbf{E}_m = \mathbf{R}\mathbf{E} \tag{5.23}$$

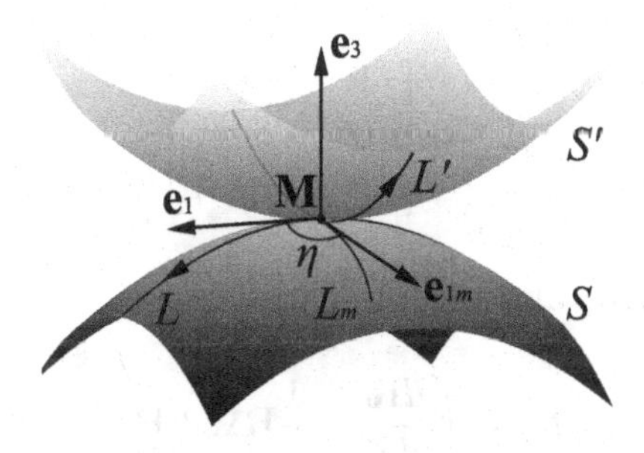

Fig. 5.3: The imprint curve L_m on the fixed surface

where

$$\mathbf{E}_m = \begin{bmatrix} \mathbf{e}_{1m} \\ \mathbf{e}_{2m} \\ \mathbf{e}_{3m} \end{bmatrix}, \quad \mathbf{R} = \begin{bmatrix} \cos\eta & \sin\eta & 0 \\ -\sin\eta & \cos\eta & 0 \\ 0 & 0 & 1 \end{bmatrix}, \quad \mathbf{E} = \begin{bmatrix} \mathbf{e}_1 \\ \mathbf{e}_2 \\ \mathbf{e}_3 \end{bmatrix}$$

The arc length of the imprint curve L_m is equal to that of the curve L' because of the rolling constraints. Differentiating the left side of Eq. (5.23) w.r.t. S' yields

$$\frac{d\mathbf{E}_m}{ds'} = \boldsymbol{\Omega}_m \mathbf{E}_m = \boldsymbol{\Omega}_m \mathbf{R}\mathbf{E} \tag{5.24}$$

where the matrix $\boldsymbol{\Omega}_m$ is the Darboux matrix that contains the geodesic curvature k_{gm}, the normal curvature k_{nm}, and the geodesic torsion τ_{gm} of the imprint curve L_{im}:

$$\boldsymbol{\Omega}_m = \begin{bmatrix} 0 & k_{gm} & k_{nm} \\ -k_{gm} & 0 & \tau_{gm} \\ -k_{nm} & -\tau_{gm} & 0 \end{bmatrix}$$

Differentiating the right side of Eq. (5.23) w.r.t. S' yields

$$\frac{d}{ds'}(\mathbf{R}\mathbf{E}) = \frac{d\mathbf{R}}{ds'}\mathbf{E} + \mathbf{R}\frac{ds}{ds'}\frac{d\mathbf{E}}{ds} = \left(\frac{d\mathbf{R}}{ds'} + \frac{1}{\lambda}\mathbf{R}\boldsymbol{\Omega}\right)\mathbf{E} \tag{5.25}$$

where

$$\boldsymbol{\Omega} = \begin{bmatrix} 0 & k_g & k_n \\ -k_g & 0 & \tau_g \\ -k_n & -\tau_g & 0 \end{bmatrix}$$

Equating Eqs. (5.25) and (5.24) yields

$$\boldsymbol{\Omega}_m = \left(\frac{d\mathbf{R}}{ds'} + \frac{1}{\lambda}\mathbf{R}\boldsymbol{\Omega}\right)\mathbf{R}^{-1} \tag{5.26}$$

This geodesic curvature k_{gm}, the normal curvature k_{nm}, and the geodesic torsion τ_{gm} of the imprint curve L_{im} can be obtained from Eq. (5.26) as

$$\begin{aligned} k_{gm} &= \frac{k_g}{\lambda} + \frac{d\eta}{ds'} \\ k_{nm} &= \frac{1}{\lambda}(k_n \cos\eta + \tau_g \sin\eta) \\ \tau_{gm} &= \frac{1}{\lambda}(-k_n \sin\eta + \tau_g \cos\eta) \end{aligned} \tag{5.27}$$

The angular velocity caused by the imprint curve L_m on the fixed surface and the contact trajectory curve L' on the moving surface can be obtained from Eq. (4.16) as

$$\begin{aligned} \omega = &-\sigma'(\tau_g' - \tau_{gm})\mathbf{e}_{1m} \\ &+\sigma'(k_n' - k_{gn})\mathbf{e}_{2m} \\ &-\sigma'(k_g' - k_{gm})\mathbf{e}_{3m} \end{aligned} \tag{5.28}$$

It can be verified that the angular velocity in Eq. (5.28) is equal to that in Eq. (5.19). We have hence clarified the causes of the angular velocity by revealing the geometric properties of the imprint curve L_m.

5.6 The Classical Problem Revisited

5.6.1 *Closed-Form Solution*

Consider a disc of unit radius rolling–sliding on a plane while maintaining upright. The contact trajectory curve on the plane is L and the contact trajectory curve on the disc is the unit circle L'. Let η represent the angle between L and L' at the contact point M, σ the rolling–sliding rate on L, and σ' rolling rate on L'.

We set up the Darboux frame ($\mathbf{M}$, $\mathbf{e}_1$, $\mathbf{e}_2$, $\mathbf{e}_3$) on L and ($\mathbf{M}$, $\mathbf{e}_1'$, $\mathbf{e}_2'$, $\mathbf{e}_3'$) on L' such that $\mathbf{e}_3$ points to the disc center and $\mathbf{e}_3 = \mathbf{e}_3'$, as shown in Fig. 5.4.

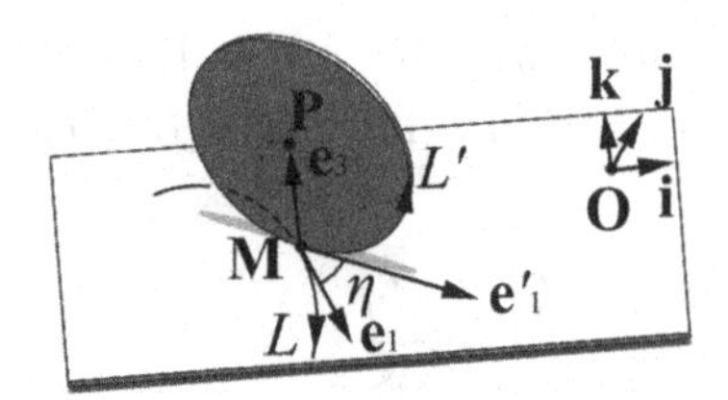

Fig. 5.4: A disc of unit radius rolling–sliding on a plane

The Darboux formulas of the curve L on the plane can be obtained as

$$\frac{d}{ds}\begin{bmatrix}\mathbf{e}_1\\ \mathbf{e}_2\\ \mathbf{e}_3\end{bmatrix} = \begin{bmatrix}0 & k & 0\\ -k & 0 & 0\\ 0 & 0 & 0\end{bmatrix}\begin{bmatrix}\mathbf{e}_1\\ \mathbf{e}_2\\ \mathbf{e}_3\end{bmatrix} \tag{5.29}$$

where k is the curvature of L at M.

The Darboux formulas of the curve L' on the disc can be obtained as

$$\frac{d}{ds}\begin{bmatrix}\mathbf{e}'_1\\ \mathbf{e}'_2\\ \mathbf{e}'_3\end{bmatrix} = \begin{bmatrix}0 & 0 & 1\\ 0 & 0 & 0\\ -1 & 0 & 0\end{bmatrix}\begin{bmatrix}\mathbf{e}'_1\\ \mathbf{e}'_2\\ \mathbf{e}'_3\end{bmatrix} \tag{5.30}$$

The translational velocity and rotational velocity of the disc can be obtained from Eq. (5.19) as

$$\begin{aligned}\mathbf{v} &= (\sigma - \sigma' \cos\eta)\mathbf{e}_1 - \sigma' \sin\eta \mathbf{e}_2\\ \boldsymbol{\omega} &= -\sigma'(\sin\eta \mathbf{e}_1 - \cos\eta \mathbf{e}_2) + \left(\frac{d\eta}{dt} + \sigma k\right)\mathbf{e}_3\end{aligned} \tag{5.31}$$

Remark 5.1. At first sight, the angular velocity in Eq. (5.31) appears to be violating the geometric constraints of the disc maintaining an upright posture, since the angular velocity in the direction of $\mathbf{e}_1$ is not 0. However, this angular velocity is referenced w.r.t. the frame ($\mathbf{M}$, $\mathbf{e}'_1$, $\mathbf{e}'_2$, $\mathbf{e}'_3$) as

$$\boldsymbol{\omega}' = \sigma'\mathbf{e}'_2 + \left(\frac{d\eta}{dt} + \sigma k\right)\mathbf{e}'_3$$

It can be seen that the angular velocity in the direction of $\mathbf{e}'_1$ is 0, thus it does not violate the constraint of the disc maintaining an upright posture.

The position vector from the point M to P referenced w.r.t. the frame ($\mathbf{M}$, $\mathbf{e}_1$, $\mathbf{e}_2$, $\mathbf{e}_3$) is $\mathbf{e}_3$. The velocity of the point P is thus

$$\mathbf{v}_P = \mathbf{v} + \boldsymbol{\omega} \times \mathbf{e}_3 = \sigma \mathbf{e}_1$$

If the direction of sliding is parallel to that of rolling, rolling–sliding contact becomes rolling–slipping and the angle η becomes 0. In this case, the velocity of the disc is

$$\begin{aligned} \mathbf{v} &= (\sigma - \sigma')\mathbf{e}_1 \\ \boldsymbol{\omega} &= \sigma' \mathbf{e}_2 + \sigma k \mathbf{e}_3 \end{aligned}$$

5.6.2 *Numerical Simulation*

Suppose the contact curve L on the plane S is a circle of curvature 0.25, the rolling–sliding rate σ is 1, the ratio of rolling rate to the sliding–rolling rate λ is 0.8, and the angle η between rolling and sliding is a constant $\pi/6$, as shown in Fig. 5.5.

Set up a fixed frame ($\mathbf{O}$, $\mathbf{i}$, $\mathbf{j}$, $\mathbf{k}$) in such a way that the $\mathbf{k}$-axis is perpendicular to the plane S, as shown in Fig. 5.5. Suppose the angle between the $\mathbf{i}$-axis and OM is θ, and the disc is at the intersection between the circle and the $\mathbf{i}$-axis at the starting time $t_0 = 0$. The angle θ can be obtained as $\theta = sk = t/4$, where $s = \sigma t$ is the arc

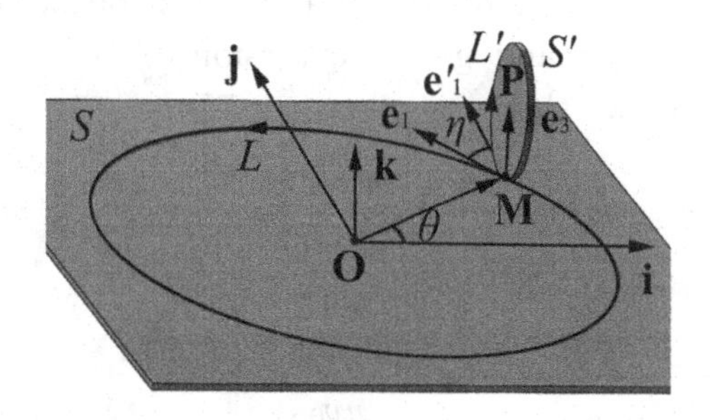

Fig. 5.5: A disc of unit radius rolling–sliding along a circle of curvature 0.25

length covered in the period of time t and k is the curvature of the curve L.

It follows that the vectors $\mathbf{e}_1$, $\mathbf{e}_2$, and $\mathbf{e}_3$ w.r.t. the fixed frame $(\mathbf{O}, \mathbf{i}, \mathbf{j}, \mathbf{k})$ are

$$\begin{aligned}\mathbf{e}_1 &= \left[-\sin\tfrac{t}{4}\ \cos\tfrac{t}{4}\ 0\right]\\ \mathbf{e}_2 &= \left[-\cos\tfrac{t}{4}\ -\sin\tfrac{t}{4}\ 0\right]\\ \mathbf{e}_3 &= \left[0\ 0\ 1\right]\end{aligned}$$

It follows Eq. (5.19) to obtain the angular velocity of the disc and the velocity of the center point P, which are plotted in Fig. 5.6.

5.7 Example of Application

Consider a sphere of radius R rolling–sliding on a paraboloid, which is formed by rotating a parabola $z = \frac{y^2}{2}$ about the $\mathbf{z}$-axis. The contact curves are a small circle L' on the sphere and a meridian L on the paraboloid, as shown in Fig. 5.7.

We set up the Darboux frame $(\mathbf{M}\text{-}\mathbf{e}_1, \mathbf{e}_2, \mathbf{e}_3)$ and $(\mathbf{M}\text{-}\mathbf{e}'_1, \mathbf{e}'_2, \mathbf{e}'_3)$ such that $\mathbf{e}_1$ is the tangent vector to L and $\mathbf{e}'_1$ the tangent vector to L', $\mathbf{e}_3$ the common normal vector of the surfaces S and S', as shown in Fig. 5.7.

The meridian L can be parameterized as

$$\mathbf{x}(u, v_0) = \left[u\ v_0\ \tfrac{1}{2}(u^2 + v_0^2)\right]$$

In Example 4.2, we have computed the geodesic curvature k_g, normal curvature k_n, and geodesic torsion τ_g of a meridian on a paraboloid. It follows that

$$\begin{aligned}k_g &= \frac{-v_0}{(u^2+1)^{\frac{3}{2}}\sqrt{u^2+v_0^2+1}}\\ k_n &= \frac{1}{(u^2+1)\sqrt{u^2+v_0^2+1}}\\ \tau_g &= \frac{uv_0}{(u^2+1)(u^2+v_0^2+1)}\end{aligned} \tag{5.32}$$

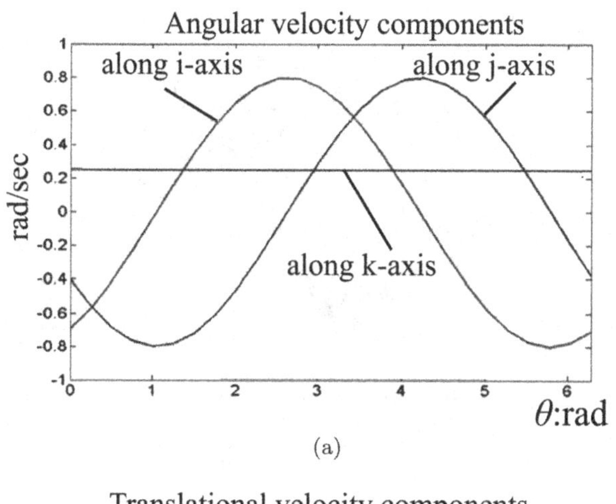

(a)

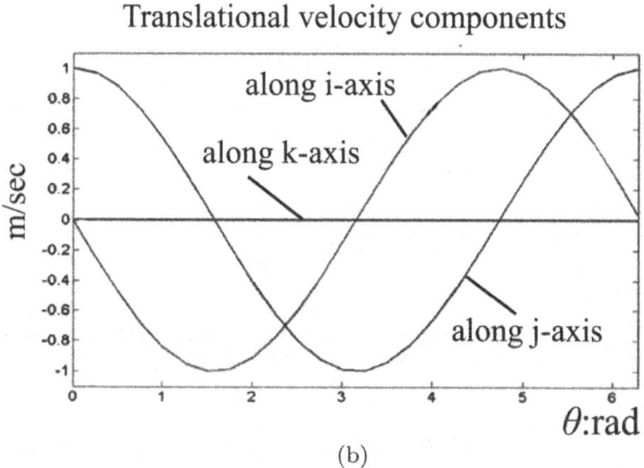

(b)

Fig. 5.6: The angular velocity of the disk and the translational velocity of the center point P: (a) angular velocity components along $\mathbf{i}$, $\mathbf{j}$, and $\mathbf{k}$-axes; (b) translational velocity components along the $\mathbf{i}$, $\mathbf{j}$, and $\mathbf{k}$-axes

The geodesic curvature k'_g, normal curvature k'_n, and geodesic torsion τ'_g of the small circle L' can be obtained as

$$\begin{aligned} k'_g &= -\frac{\cot\delta}{R} \\ k'_n &= \frac{1}{R} \\ \tau'_g &= 0 \end{aligned} \tag{5.33}$$

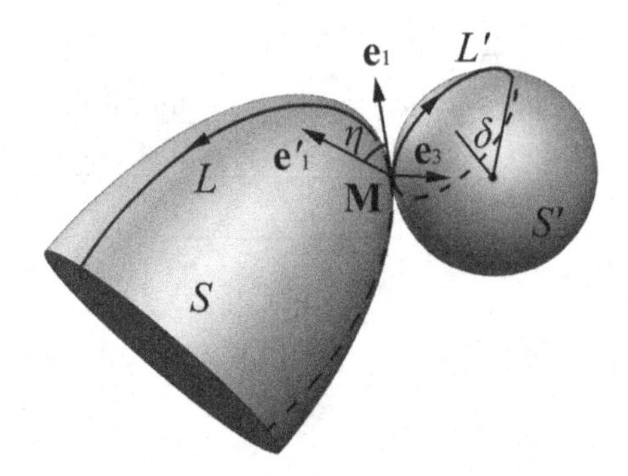

Fig. 5.7: A sphere rolling–sliding on a paraboloid

where δ is the half cone-angle of the small circle L, as shown in Fig. 5.7.

The angular velocity and translational velocity of the sphere can be obtained by substituting Eqs. (5.32) and (5.33) into Eq. (5.19):

$$\begin{aligned}
\mathbf{v} &= \sigma(1-\lambda)\mathbf{e}_1 - \sigma\lambda\mathbf{e}_2 \\
\boldsymbol{\omega} &= \left(-\frac{\sigma\lambda}{R}\sin\eta + \frac{\sigma u v_0}{(u^2+1)(u^2+v_0^2+1)}\right)\mathbf{e}_1 \\
&\quad + \left(\frac{\sigma\lambda}{R}\cos\eta - \frac{\sigma}{(u^2+1)\sqrt{u^2+v_0^2+1}}\right)\mathbf{e}_2 \\
&\quad + \left(-\frac{\sigma\lambda\cot\delta}{R}\sin\eta + \frac{d\eta}{dt} + \frac{-\sigma v_0}{(u^2+1)^{\frac{3}{2}}\sqrt{u^2+v_0^2+1}}\right)\mathbf{e}_3
\end{aligned}$$

5.8 Conclusion

We have solved the forward kinematics of rolling–sliding contact when the two contact curves do not have a common tangent vector. The kinematics of rolling–slipping contact can be deduced as a special case. We have further revealed the properties of the instantaneous imprint curve and validated them through the formulation of the angular velocity. Finally, we have revisited a classical problem and given an example of two surfaces with variable curvatures.

Part III

Inverse Kinematics of Rolling–Sliding Contact

Chapter 6

From Velocity to Trajectories: Inverse Kinematics of Rigid Surfaces with Rolling Contact

In this chapter, we consider two surfaces subject to rolling contact and obtain the contact trajectories when the angular velocity of the moving surface is present. We obtain three scalar entities that generate the contact trajectories to realize the given angular velocity of the moving surface: the rolling direction φ, the rolling rate σ, and the complementary spin rate ω_{spin}.

6.1 Problem Definition

Suppose the fixed surface S is parameterized as $\mathbf{x}(u, v)$. Let $\mathbf{e}_u$ denote the unit tangent vector of the u-curve and $\mathbf{e}_v$ the unit tangent vector of the v-curve, as shown in Fig. 6.1.

Suppose the moving surface S' rolls on the fixed surface S. The angular velocity is given w.r.t. the frame ($\mathbf{M}$, $\mathbf{e}_u$, $\mathbf{e}_v$, $\mathbf{e}_3$) as

$$\boldsymbol{\omega} = \omega_1\mathbf{e}_u + \omega_2\mathbf{e}_v + \omega_3\mathbf{e}_3 \tag{6.1}$$

We obtain the rolling direction φ, the rolling rate σ, and the complementary spin rate ω_{spin}, which generate the contact trajectories to realize the given angular velocity in Eq. (6.1).

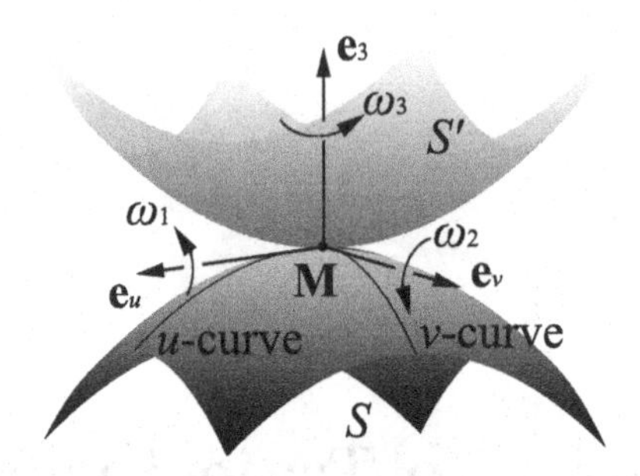

Fig. 6.1: The desired angular velocity referenced w.r.t. the frame ($\mathbf{M}$, $\mathbf{e}_u$, $\mathbf{e}_v$, $\mathbf{e}_3$)

6.2 The Angular Velocity in Relation to Coordinate Curves

Suppose the moving surface S' parameterized as $\mathbf{x}'(\alpha, \beta)$. Let $\mathbf{e}_\alpha$ denote the unit tangent vector along the α-curve, M denote the contact point, and θ the angle between $\mathbf{e}_u$ and $\mathbf{e}_\alpha$.

Suppose there exits a contact trajectory L on S and a contact trajectory L' on S' that can generate the given angular velocity in Eq. (6.1) and the common tangent vector $\mathbf{e}_1$ of these two contact trajectories makes an angle φ with $\mathbf{e}_\alpha$, as shown in Fig. 6.2.

Let s denote arc length of the contact trajectory L on the fixed surface S. The geodesic curvature k_g, normal curvature k_n, and geodesic torsion τ_g of L can be obtained from Eq. (2.62) as

$$\begin{aligned} k_g &= k_{gu}\cos(\varphi+\theta) + k_{gv}\sin(\varphi+\theta) + \frac{d(\varphi+\theta)}{ds} \\ k_n &= k_{nu}\cos^2(\varphi+\theta) + 2\tau_{gu}\cos(\varphi+\theta)\sin(\varphi+\theta) + k_{nv}\sin^2(\varphi+\theta) \\ \tau_g &= \tau_{gu}\cos 2(\varphi+\theta) + \frac{1}{2}(k_{nv}-k_{nu})\sin 2(\varphi+\theta) \end{aligned} \tag{6.2}$$

where k_{gu}, k_{gv}, k_{nu}, k_{nv}, τ_{gu}, and τ_{gv} are the curvatures of the u-curve and v-curve of S.

Let s' denote arc length of the contact trajectory L' on the moving surface S'. The geodesic curvature k'_g, normal curvature k'_n, and

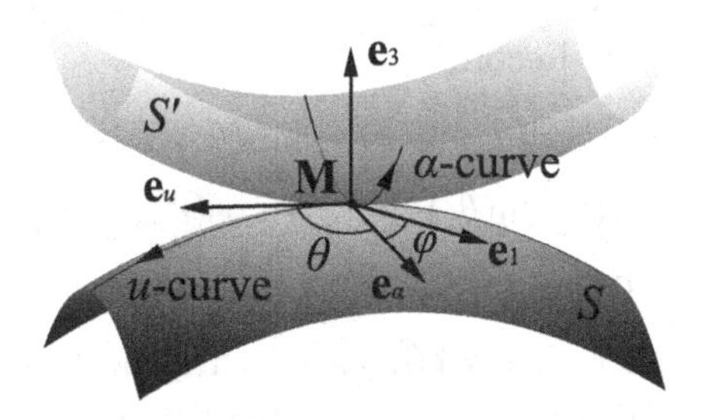

Fig. 6.2: The rolling direction and coordinate curves

geodesic torsion τ_g' can be obtained from Eq. (2.62) as

$$\begin{aligned}
k_g' &= k_{g\alpha} \cos\varphi + k_{g\beta} \sin\varphi + \frac{d\varphi}{ds'} \\
k_n' &= k_{n\alpha} \cos^2\varphi + 2\tau_{g\alpha} \cos\varphi \sin\varphi + k_{n\beta} \sin^2\theta \\
\tau_g' &= \tau_{g\alpha} \cos 2\varphi + \frac{1}{2}(k_{n\beta} - k_{n\alpha}) \sin 2\varphi
\end{aligned} \tag{6.3}$$

where $k_{g\alpha}$, $k_{g\beta}$, $k_{n\alpha}$, $k_{n\beta}$, $\tau_{g\alpha}$, and $\tau_{g\beta}$ are the curvatures of the α-curve and β-curve of S'.

When the two contact trajectories are present, the angular velocity of the moving object can be obtained from Eq. (4.16) as

$$\boldsymbol{\omega} = \sigma(-\tau_g^* \mathbf{e}_1 + k_n^* \mathbf{e}_2 - k_g^* \mathbf{e}_3) \tag{6.4}$$

where

$$\sigma = \frac{ds}{dt}, \ \tau_g^* = \tau_g' - \tau_g, \ k_n^* = k_n' - k_n, \ k_g^* = k_g' - k_g$$

The frame ($\mathbf{M}$, $\mathbf{e}_1$, $\mathbf{e}_2$, $\mathbf{e}_3$) can be obtained from ($\mathbf{M}$, $\mathbf{e}_u$, $\mathbf{e}_v$, $\mathbf{e}_3$) by rotating an angle $(\theta + \varphi)$ about the $\mathbf{e}_3$ direction. It follows that

$$\begin{aligned}
\mathbf{e}_1 &= \cos(\theta + \varphi)\mathbf{e}_u + \sin(\theta + \varphi)\mathbf{e}_v \\
\mathbf{e}_2 &= -\sin(\theta + \varphi)\mathbf{e}_u + \cos(\theta + \varphi)\mathbf{e}_v \\
\mathbf{e}_3 &= \mathbf{e}_3
\end{aligned} \tag{6.5}$$

When referenced w.r.t. the frame ($\mathbf{M}$, $\mathbf{e}_u$, $\mathbf{e}_v$, $\mathbf{e}_3$), the angular velocity $\boldsymbol{\omega}$ in Eq. (6.4) becomes

$$\begin{aligned}\boldsymbol{\omega} &= \sigma(-\cos(\theta+\varphi)\tau_g^* - \sin(\theta+\varphi)k_n^*)\mathbf{e}_u \\ &\quad + \sigma(-\sin(\theta+\varphi)\tau_g^* + \cos(\theta+\varphi)k_n^*)\mathbf{e}_v \\ &\quad - \sigma k_g^* \mathbf{e}_3\end{aligned} \tag{6.6}$$

Equating Eqs. (6.1) and (6.6) yields three scalar equations along the $\mathbf{e}_u$, $\mathbf{e}_v$, and $\mathbf{e}_3$ directions. If the values of ω_i are given, the three scalar contact equations form a system of nonlinear equations with three variables: σ, φ, and $d\varphi/ds$. Each of the three variables has distinct physical meaning: σ representing the rolling rate, φ representing the rolling direction, and $d\varphi/ds$ representing the spin rate about the normal.

Substituting the curvatures in Eqs. (6.2) and (6.3) into Eq. (6.6) yields the three contact equations:

$$\begin{aligned}\omega_1 &= \sigma(A_1\cos\varphi + A_2\sin\varphi) \\ \omega_2 &= \sigma(B_1\cos\varphi + B_2\sin\varphi) \\ \omega_3 &= \sigma(C_1\cos\varphi + C_2\sin\varphi) + \omega_{\text{spin}}\end{aligned} \tag{6.7}$$

where

$$\begin{aligned}A_1 &= \tau_{gu}\cos\theta - \tau_{g\alpha}\cos\theta - k_{n\alpha}\sin\theta + k_{nv}\sin\theta \\ A_2 &= k_{nv}\cos\theta - k_{n\beta}\cos\theta - \tau_{g\alpha}\sin\theta - \tau_{gu}\sin\theta \\ B_1 &= k_{n\alpha}\cos\theta - k_{nu}\cos\theta - \tau_{g\alpha}\sin\theta - \tau_{gu}\sin\theta \\ B_2 &= \tau_{g\alpha}\cos\theta - \tau_{gu}\cos\theta - k_{n\beta}\sin\theta + k_{nu}\sin\theta \\ C_1 &= k_{gu}\cos\theta + k_{gv}\sin\theta - k_{g\alpha} \\ C_2 &= k_{gv}\cos\theta - k_{gu}\sin\theta - k_{g\beta} \\ \omega_{\text{spin}} &= \sigma\frac{d\theta}{ds} = \frac{ds}{dt}\frac{d\theta}{ds} = \frac{d\theta}{dt}\end{aligned}$$

Note the coefficients A_i, B_i, and C_i are known at the given contact point M.

Remark 6.1. The third contact equation in Eq. (6.7) represents the spin rate about the normal $\mathbf{e}_3$. The first two items in this equation, $\sigma(C_1\cos\varphi + C_2\sin\varphi)$, represent the spin rate induced by the geometry of the two surfaces. The last item, ω_{spin}, is the difference between the desired spin rate and the induced spin rate. We named it the **compensatory spin rate**.

Eliminating σ from the first two equations in Eq. (6.7) yields

$$(B_1\omega_1 - A_1\omega_2)\cos\varphi + (B_2\omega_1 - A_2\omega_2)\sin\varphi = 0 \tag{6.8}$$

The rolling direction φ can be obtained from Eq. (5.8) as

$$\varphi = \arctan\frac{A_1\omega_2 - B_1\omega_1}{B_2\omega_1 - A_2\omega_2} \tag{6.9}$$

The rolling rate σ and compensatory spin rate ω_{spin} can be straightforwardly obtained from Eq. (6.7) when the rolling direction φ is known.

6.3 Reconciliation with a Classical Example: A Unit Ball Rolling on a Plane

6.3.1 *A Closed-Form Solution*

We revisit a classical example — a unit ball rolling on a plane — to show the reconciliation of Eq. (6.7) with the classical solution.

Consider a unit sphere S' rolling on a fixed plane S, as shown in Fig. 6.3.

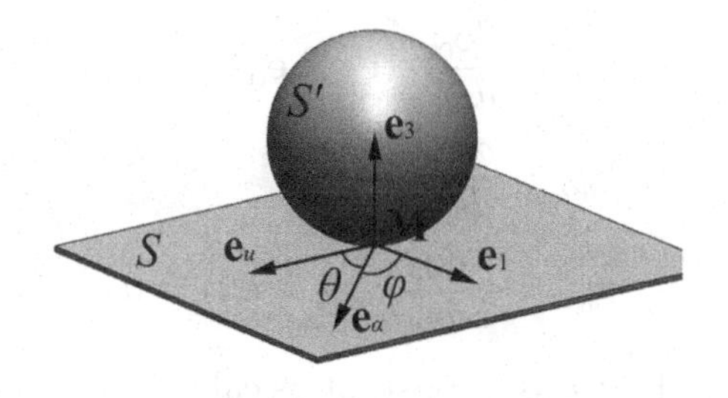

Fig. 6.3: A unit sphere rolling on a plane

The plane S can be parameterized as

$$\mathbf{x}(u,v) = \left[u, \, , v, 0\right]$$

Let the normal $\mathbf{e}_3$ point upwards. Both the u- and v-curves are straight-lines. Hence, the geodesic curvature, normal curvature, and geodesic torsion of the u- and v-curves are 0. It follows that

$$k_{gu} = k_{nu} = \tau_{gu} = k_{gv} = k_{nv} = \tau_{gv} = 0 \tag{6.10}$$

The unit sphere S' can be parameterized as

$$\mathbf{x}'(\alpha,\beta) = \left[\cos\alpha\cos\beta, \, -\cos\alpha\sin\beta, \, \sin\alpha\right]$$
$$\alpha \in (-\pi/2, \pi/2), \beta \in [0, 2\pi)$$

The α-curve, where β is a constant, is called meridians. Let s_α denote the arc length of the α-curve. The infinitesimal arc length of the α-curve can be obtained as

$$\frac{ds_\alpha}{d\alpha} = \left|\frac{d\mathbf{x}'}{d\alpha}\right| = 1$$

The unit tangent vector $\mathbf{e}_\alpha$ can be obtained as

$$\mathbf{e}_\alpha = \frac{\frac{d\mathbf{x}'}{d\alpha}}{\left|\frac{d\mathbf{x}'}{d\alpha}\right|} = \left[-\sin\alpha\cos\beta, \, \sin\alpha\sin\beta, \, \cos\alpha\right]$$

The normal $\mathbf{e}_3$ points toward the sphere center. It follows that

$$\mathbf{e}_3 = -\mathbf{x}'(\alpha,\beta) = -\left[\cos\alpha\cos\beta, \, -\cos\alpha\sin\beta, \, \sin\alpha\right]$$

The geodesic curvature $k_{g\alpha}$, normal curvature $k_{n\alpha}$, and geodesic torsion $\tau_{g\alpha}$ can be obtained as

$$\begin{aligned} k_{g\alpha} &= \frac{d\mathbf{e}_\alpha}{ds_\alpha} \cdot (\mathbf{e}_3 \times \mathbf{e}_\alpha) = 0 \\ k_{n\alpha} &= \frac{d\mathbf{e}_\alpha}{ds_\alpha} \cdot \mathbf{e}_3 = 1 \\ \tau_{g\alpha} &= -\frac{d\mathbf{e}_3}{ds_\alpha} \cdot (\mathbf{e}_3 \times \mathbf{e}_\alpha) = 0 \end{aligned} \tag{6.11}$$

The β-curve, where α is a constant, is called parallel. Let s_β denote the arc length of the β-curve. The infinitesimal arc length of the

β-curve can be obtained as

$$\frac{ds_\beta}{d\beta} = \left|\frac{d\mathbf{x}'}{d\beta}\right| = \cos\alpha$$

The unit tangent vector $\mathbf{e}_\beta$ can be obtained as

$$\mathbf{e}_\beta = \frac{\frac{d\mathbf{x}'}{d\beta}}{\left|\frac{d\mathbf{x}'}{d\beta}\right|} = [-\sin\beta, -\cos\beta, 0]$$

The geodesic curvature $k_{g\beta}$, normal curvature $k_{n\beta}$, and geodesic torsion $\tau_{g\beta}$ of the β-curve can be obtained as

$$\begin{aligned}
k_{g\beta} &= \frac{d\mathbf{e}_\beta}{ds_\beta} \cdot (\mathbf{e}_3 \times \mathbf{e}_\beta) = \tan\alpha \\
k_{n\beta} &= \frac{d\mathbf{e}_\beta}{ds_\beta} \cdot \mathbf{e}_3 = 1 \\
\tau_{g\beta} &= -\frac{d\mathbf{e}_3}{ds_\beta} \cdot (\mathbf{e}_3 \times \mathbf{e}_\beta) = 0
\end{aligned} \tag{6.12}$$

Let $\mathbf{e}_u$ represent the unit tangent vector to the u-curve of S at the contact point M and θ the angle between $\mathbf{e}_u$ and $\mathbf{e}_\alpha$.

Suppose the sphere rolls on the plane in the direction of $\mathbf{e}_1$, which makes an angle φ with $\mathbf{e}_\alpha$, as shown in Fig. 6.3, and its angular velocity w.r.t. the frame $(\mathbf{M}, \mathbf{e}_u, \mathbf{e}_v, \mathbf{e}_3)$ is given as

$$\boldsymbol{\omega} = \omega_1\mathbf{e}_u + \omega_2\mathbf{e}_v + \omega_3\mathbf{e}_3$$

Substituting Eqs. (6.11) and (6.12) into Eq. (6.7) yields three contact equations as

$$\begin{aligned}
\omega_1 &= -\sigma\sin(\theta + \varphi) \\
\omega_2 &= \sigma\cos(\theta + \varphi) \\
\omega_3 &= -\sigma\tan\alpha\sin\varphi + \omega_{\text{spin}}
\end{aligned} \tag{6.13}$$

Solving Eq. (6.13) for the three variables φ, θ, and ω_{spin} gives

$$\begin{aligned} \varphi &= -\tan^{-1}\left(\frac{\omega_1}{\omega_2}\right) - \theta \\ \sigma &= \sqrt{\omega_1^2 + \omega_2^2} \\ \omega_{\text{spin}} &= \omega_3 + \sigma \tan\alpha \sin\varphi \end{aligned} \tag{6.14}$$

Note that the third variable ω_{spin} contains two entities: The first one, ω_3, is the desired spin rate. The second, $\sigma \tan\alpha \sin\varphi$, is the spin rate induced by the geometry of the two surfaces.

6.3.2 *The Classical Solution: A System of Differential Equations*

In the classical approaches, for example [Montana, 1988], the contact trajectories are expressed as a nonlinear system of five first-order differential equations, which is developed from the constraints of rolling contact. The unit-ball-rolling-on-a-plane problem is formulated as the following five differential equations after $\mathbf{e}_u$ and $\mathbf{e}_\alpha$ are assumed to coincide:

$$\begin{aligned} \dot{u} &= -\omega_2 \\ \dot{v} &= \omega_1 \\ \dot{\alpha} &= -\omega_2 \\ \dot{\beta} &= -\omega_1 \sec\alpha \\ \dot{\theta} &= \omega_3 - \omega_1 \tan\alpha \end{aligned} \tag{6.15}$$

where a dot over a variable indicates a derivative taken w.r.t. time t.

We show Eqs. (6.14) and (6.15) are equivalent. First, the assumption about $\mathbf{e}_u$ and $\mathbf{e}_\alpha$ coinciding in the classical solution gives $\theta = 0$ in Eq. (6.14). Second, the rolling rate in Eq. (6.14) is defined as $\sigma = ds/dt$, but $ds/dt = \sqrt{\dot{u}^2 + \dot{v}^2}$, and subsequently equal to $\sqrt{\omega_1^2 + \omega_2^2}$ from the first two equations in Eq. (6.15). Third, the rolling direction φ in Eq. (6.14) can be obtained as $\varphi = \arctan(\dot{v}/\dot{u}) = -\arctan(\omega_2/\omega_1)$ from Eq. (6.15), as shown in Fig. 6.4.

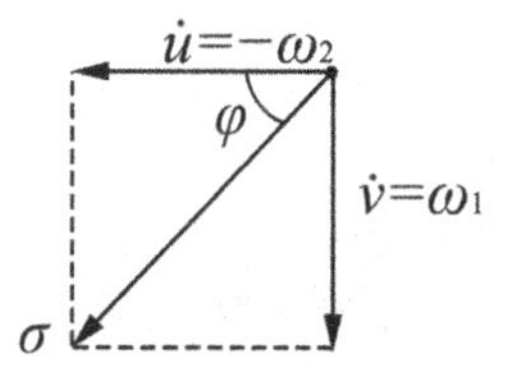

Fig. 6.4: The geometry of the rolling direction and rolling rate

Finally, the compensatory spin rate ω_{spin} in Eq. (6.14) is equal to the last equation in Eq. (6.15) since $\sigma \sin \varphi = -\omega_1$.

6.4 Examples of Application

6.4.1 *An Ellipsoid Rolling on a Plane*

Consider an ellipsoid rolling on a plane, as shown in Fig. 6.5.

We obtain a closed-form solution to the inverse kinematics when the angular velocity of the ellipsoid is given as

$$\boldsymbol{\omega} = \omega_1 \mathbf{e}_u + \omega_2 \mathbf{e}_v + \omega_3 \mathbf{e}_3$$

An ellipsoid, also called a triaxial ellipsoid, is a quadratic surface, which can be parameterized as

$$\mathbf{x}'(\alpha, \beta) = \left[b \cos \beta,\, a \cos \alpha \sin \beta,\, a \sin \alpha \sin \beta\right]$$
$$\alpha \in [0, 2\pi), \beta \in (0, \pi)$$

Let the normal $\mathbf{e}_3$ point upward. We have known that the geodesic curvature, normal curvature, geodesic torsion of the u- and v-curves of the plane are 0. The geodesic curvature, normal curvature, geodesic torsion of the α- and β-curves of the ellipsoid can be obtained as

$$k_{g\alpha} = \frac{-\cos \beta}{D \sin \beta}, \quad k_{n\alpha} = \frac{-b}{aD}, \quad \tau_{g\alpha} = 0$$
$$k_{g\beta} = 0, \quad k_{n\beta} = \frac{-ab}{D^3}, \quad \tau_{g\beta} = 0$$

where $D = \sqrt{a^2 \cos^2 \beta + b^2 \sin^2 \beta}$.

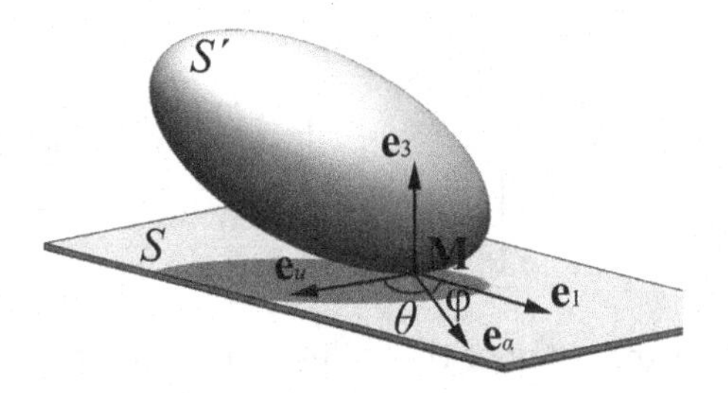

Fig. 6.5: An ellipsoid rolling on a plane

The three contact equations can be obtained from Eq. (6.7) as

$$\begin{aligned}\omega_1 &= \frac{\sigma b}{aD^3}(a^2\cos\theta\sin\varphi + D^2\sin\theta\cos\varphi)\\ \omega_2 &= \frac{\sigma}{aD^3}(a^2 b\sin\theta\sin\varphi - bD^2\cos\theta\cos\varphi)\\ \omega_3 &= \frac{\sigma\cos\beta\cos\varphi}{D\sin\beta} + \omega_{\text{spin}}\end{aligned} \tag{6.16}$$

The rolling direction φ can be obtained from the first two equations of Eq. (6.16) as

$$\varphi = -\arctan\left(\frac{D^2(\omega_1\cos\theta + \omega_2\sin\theta)}{a^2(\omega_2\cos\theta - \omega_1\sin\theta)}\right)$$

The rolling rate σ and compensatory spin rate ω_{spin} can be obtained by substituting φ into Eq. (6.16). □

6.4.2 *A Unit Sphere Rolling on a Paraboloid*

Consider a unit sphere rolling on a paraboloid, as shown in Fig. 6.6.

We obtain a closed-form solution to the inverse kinematics when the angular velocity of the sphere is given as

$$\boldsymbol{\omega} = \omega_1\mathbf{e}_u + \omega_2\mathbf{e}_v + \omega_3\mathbf{e}_3$$

The paraboloid can be parameterized as

$$\mathbf{x}(u,v) = \left[u\cos v,\ ,u\sin v,\ \tfrac{u^2}{2}\right]$$

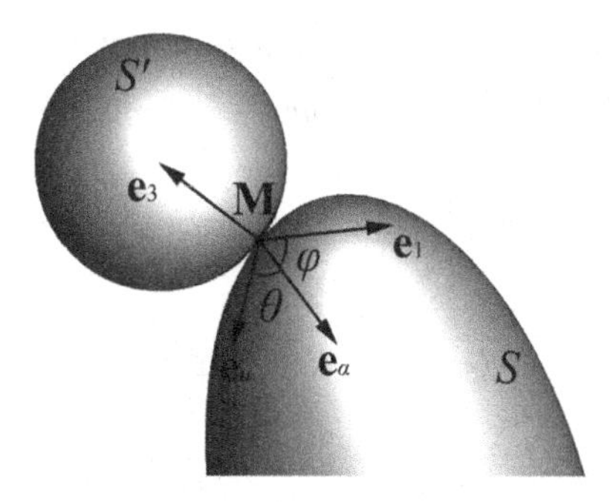

Fig. 6.6: A unit sphere rolling on a paraboloid

Let s_u denote the arc length of the u-curve, where v is a constant. The infinitesimal arc length of the u-curve can be obtained as

$$ds_u = \left|\frac{d\mathbf{x}'}{du}\right| du = \sqrt{u^2+1}du$$

The unit tangent vector $\mathbf{e}_u$ can be obtained as

$$\mathbf{e}_u = \frac{\frac{d\mathbf{x}'}{du}}{|\frac{d\mathbf{x}'}{du}|} = \frac{1}{\sqrt{u^2+1}}\left[\cos v,\ \sin v,\ u\right]$$

Let the normal $\mathbf{e}_3$ points toward the sphere center. It follows that

$$\mathbf{e}_3 = \frac{\mathbf{x}_u \times \mathbf{x}_v}{|\mathbf{x}_u \times \mathbf{x}_v|} = \frac{1}{\sqrt{u^2+1}}\left[-u\cos v,\ -u\sin v,\ 1\right]$$

The geodesic curvature k_{gu}, normal curvature k_{nu}, and geodesic torsion τ_{gu} can be obtained as

$$\begin{aligned}
k_{gu} &= \frac{d\mathbf{e}_u}{ds_u} \cdot (\mathbf{e}_3 \times \mathbf{e}_u) = 0 \\
k_{nu} &= \frac{d\mathbf{e}_u}{ds_u} \cdot \mathbf{e}_3 = \frac{1}{(u^2+1)^{\frac{3}{2}}} \\
\tau_{gu} &= -\frac{d\mathbf{e}_3}{ds_u} \cdot (\mathbf{e}_3 \times \mathbf{e}_u) = 0
\end{aligned} \tag{6.17}$$

Similarly, we can obtain the geodesic curvature k_{gv}, normal curvature k_{nv}, and geodesic torsion τ_{gv} of the v-curve as

$$\begin{aligned} k_{gv} &= \frac{d\mathbf{e}_v}{ds_v} \cdot (\mathbf{e}_3 \times \mathbf{e}_v) = \frac{1}{u\sqrt{u^2+1}} \\ k_{nv} &= \frac{d\mathbf{e}_v}{ds_v} \cdot \mathbf{e}_3 = \frac{1}{\sqrt{u^2+1}} \\ \tau_{gv} &= -\frac{d\mathbf{e}_3}{ds_v} \cdot (\mathbf{e}_3 \times \mathbf{e}_v) = 0 \end{aligned} \tag{6.18}$$

Let θ denote the angle between $\mathbf{e}_u$ and $\mathbf{e}_\alpha$ at the contact point M. Suppose the sphere rolls on the plane in the direction of $\mathbf{e}_1$, which makes an angle φ with $\mathbf{e}_\alpha$, as shown in Fig. 6.6, and its angular velocity w.r.t. the frame $(\mathbf{M}, \mathbf{e}_u, \mathbf{e}_v, \mathbf{e}_3)$ is given as

$$\boldsymbol{\omega} = \omega_1 \mathbf{e}_u + \omega_2 \mathbf{e}_v + \omega_3 \mathbf{e}_3$$

We have obtained the geodesic curvature, normal curvature, and geodesic torsion of the α- and β-curves of the unit sphere in Eqs. (6.11) and (6.12). Substituting Eqs. (6.11), (6.12), (6.17), and (6.18) into Eq. (6.7) yields three contact equations as

$$\begin{aligned} \omega_1 &= -\frac{\sigma \sin(\theta+\varphi)(\sqrt{u^2+1}-1)}{\sqrt{u^2+1}} \\ \omega_2 &= \frac{\sigma \cos(\theta+\varphi)((u^2+1)^{\frac{3}{2}}-1)}{(u^2+1)^{\frac{3}{2}}} \\ \omega_3 &= -\sigma\left(\sin\varphi\left(\tan\alpha - \frac{\cos\theta}{u\sqrt{u^2+1}}\right) - \frac{\cos\varphi\sin\theta}{u\sqrt{u^2+1}}\right) + \omega_{\text{spin}} \end{aligned} \tag{6.19}$$

The rolling direction φ can be obtained from the first two equations of Eq. (6.19) as

$$\varphi = -\arctan\frac{\omega_1(D^2+D+1)}{\omega_2 D^2} - \theta$$

where $D = \sqrt{u^2+1}$. The rolling rate σ and compensatory spin rate ω_{spin} can be obtained by substituting φ into Eq. (6.19).

6.5 Conclusion

We have obtained the closed-form inverse kinematics of two surfaces maintaining rolling contact when the angular velocity of the moving surface is given. The three contact equations form a nonlinear system with three variables, each of which has explicit physical meaning. We have clarified that the spin velocity consists of two parts: one is induced by the geometry of the two surfaces and the other has to be provided externally, which is named as the compensatory spin rate. We have demonstrated the approach's reconciliation with a classical problem and provided two examples of application.

Chapter 7

From Velocity to Trajectories: Inverse Kinematics of Rigid Surfaces with Rolling–Sliding Contact

In this chapter, we extend the approach presented in Chapter 6 to rolling–sliding contact: the moving surface rolls and slides simultaneously on the fixed surface. We obtain the contact trajectories when the translational velocity and rotational velocity of the moving surface are given. We obtain five scalar entities: the sliding rate ρ, sliding direction δ, rolling direction φ, the rolling rate σ, and the complementary spin rate ω_{spin}. These five entities generate the contact trajectories to realize the given angular velocity and translational velocity of the moving surface.

7.1 Problem Definition

Suppose the fixed surface S parameterized as $\mathbf{x}(u, v)$, and let $\mathbf{e}_u$ denote the unit tangent vector of the u-curve and $\mathbf{e}_v$ the unit tangent vector of the v-curve, as shown in Fig. 7.1.

Suppose the moving surface S' rolling–sliding on the fixed surface S. The translational velocity and angular velocity of S' are given

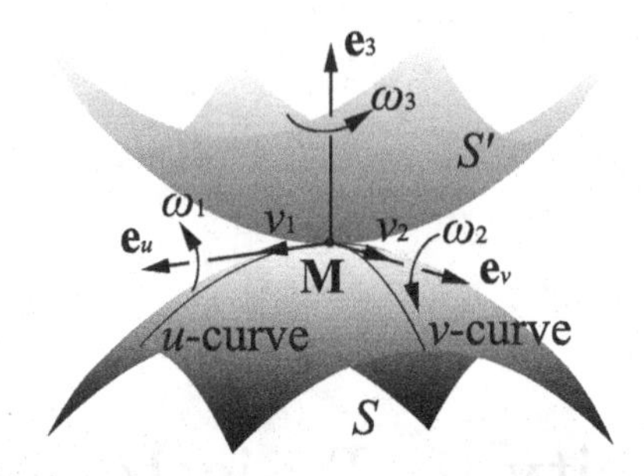

Fig. 7.1: The desired translational velocity and angular velocity referenced w.r.t. the frame ($\mathbf{M}$, $\mathbf{e}_u$, $\mathbf{e}_v$, $\mathbf{e}_3$)

w.r.t. the frame ($\mathbf{M}$, $\mathbf{e}_u$, $\mathbf{e}_v$, $\mathbf{e}_3$) as

$$\begin{aligned} \mathbf{v} &= v_1\mathbf{e}_u + v_2\mathbf{e}_v \\ \boldsymbol{\omega} &= \omega_1\mathbf{e}_u + \omega_2\mathbf{e}_v + \omega_3\mathbf{e}_3 \end{aligned} \tag{7.1}$$

We obtain the sliding rate ρ, sliding direction δ, rolling direction φ, the rolling rate σ, and the complementary spin rate ω_{spin}, which generate the contact trajectories to realize the given angular velocity in Eq. (7.1).

7.2 Sliding Rate and Sliding Direction

Suppose the moving surface S' slides at the rate of ρ in the direction of $\mathbf{e}_{1s}$, which makes an angle δ with $\mathbf{e}_u$, as shown in Fig. 7.2.

The sliding velocity of the contact point M can be obtained as

$$\mathbf{v} = \rho\mathbf{e}_{1s} = \rho\cos\delta\mathbf{e}_u + \rho\sin\delta\mathbf{e}_v \tag{7.2}$$

The rolling rate ρ and rolling direction δ can be obtained by equating Eq. (7.2) and the translational velocity in Eq. (7.1) as

$$\rho = \sqrt{v_1^2 + v_2^2},\ \delta = \arctan\frac{v_2}{v_1} \tag{7.3}$$

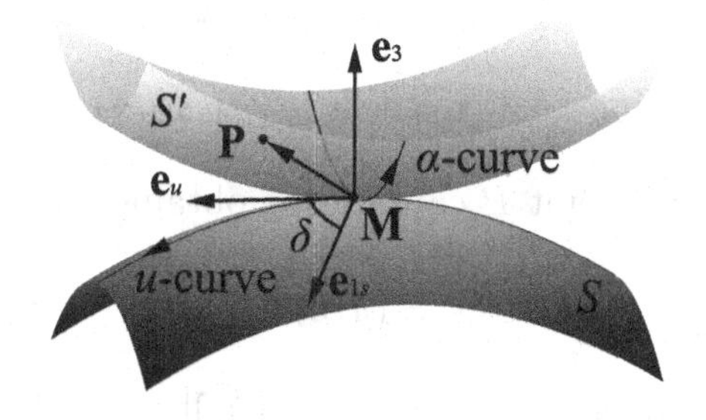

Fig. 7.2: The sliding motion in the direction of $\mathbf{e}_{1s}$

7.3 Induced Angular Velocity by the Sliding Motion and the Curved Surface

The sliding motion induces an angular velocity $\boldsymbol{\omega}_s$ due to the curved surface S. Let s_s denote the arc length of the contact trajectory curve generated by the sliding motion, and it follows that $ds_s = \rho dt$.

We set up the Darboux frame ($\mathbf{M}$, $\mathbf{e}_{1s}$, $\mathbf{e}_{2s}$, $\mathbf{e}_3$) at the contact point M in the usual way. The geodesic curvature k_{gs}, normal curvature k_{ns}, and geodesic torsion τ_{gs} in the direction of $\mathbf{e}_{1s}$ can be obtained from Eq. (2.62) straightforwardly.

Let P denote an arbitrary point on the surface S', then the position vector P referenced w.r.t. the frame ($\mathbf{O}$, $\mathbf{i}$, $\mathbf{j}$, $\mathbf{k}$) on S is

$$\mathbf{P} = \mathbf{M} + u_1\mathbf{e}_{1s} + u_2\mathbf{e}_{2s} + u_3\mathbf{e}_3 \tag{7.4}$$

where u_1 to u_3 are the coordinates of the point P referenced w.r.t. the frame ($\mathbf{M}$, $\mathbf{e}_{1s}$, $\mathbf{e}_{2s}$, $\mathbf{e}_3$). Since P is a fixed point w.r.t. the frame ($\mathbf{M}$, $\mathbf{e}_{1s}$, $\mathbf{e}_{2s}$, $\mathbf{e}_3$), it follows that each u_i is a constant and subsequently $du_i/dt = 0$. Differentiating both sides of Eq. (7.4) w.r.t. time t yields

$$\begin{aligned}\frac{d\mathbf{P}}{dt} &= \frac{ds_s}{dt}\frac{d\mathbf{P}}{ds_s} = \rho\mathbf{e}_{1s} + \rho u_1(k_{gs}\mathbf{e}_{2s} + k_{ns}\mathbf{e}_3)\\ &\quad + \rho u_2(-k_{gs}\mathbf{e}_{1s} + \tau_{gs}\mathbf{e}_3) + \rho u_3(-k_{ns}\mathbf{e}_{1s} - \tau_{gs}\mathbf{e}_{2s})\\ &= \rho(1 - u_2k_{gs} - u_3k_{ns})\mathbf{e}_{1s} + \rho(u_1k_{gs} - u_3\tau_{gs})\mathbf{e}_{2s}\\ &\quad + \rho(u_1k_{ns} + u_2\tau_{gs})\mathbf{e}_3\end{aligned} \tag{7.5}$$

Assume the angular velocity induced by the sliding motion is

$$\boldsymbol{\omega}_s = \omega_{1s}\mathbf{e}_{1s} + \omega_{2s}\mathbf{e}_{2s} + \omega_{3s}\mathbf{e}_3$$

The velocity of the point P can also be obtained as

$$\frac{d\mathbf{P}}{dt} = \rho\mathbf{e}_{1s} + \boldsymbol{\omega}_s \times (u_1\mathbf{e}_1 + u_2\mathbf{e}_2 s + u_3\mathbf{e}_3) \tag{7.6}$$

Equating Eqs. (7.6) and (7.5) yields the following three scalar equations:

$$\begin{aligned} u_2(-\omega_{3s} + \rho k_{gs}) + u_3(\omega_{2s} + \rho k_{ns}) &= 0 \\ u_1(\omega_{3s} - \rho k_{gs}) + u_3(-\omega_{1s} + \rho\tau_{gs}) &= 0 \\ u_1(-\omega_{2s} - \rho k_{ns}) + u_2(\omega_{1s} - \rho\tau_{ns}) &= 0 \end{aligned} \tag{7.7}$$

Solving Eq. (7.7) for ω_{1s}, ω_{2s}, and ω_{3s} yields

$$\begin{aligned} \omega_{1s} &= \rho\tau_{gs} \\ \omega_{2s} &= -\rho k_{ns} \\ \omega_{3s} &= \rho k_{gs} \end{aligned} \tag{7.8}$$

The angular velocity $\boldsymbol{\omega}_s$ induced by the sliding motion on the curved surface S has thus been obtained.

7.4 Inverse Kinematics of Rolling–Sliding Contact

In rolling–sliding contact, the sliding motion generates the angular velocity $\boldsymbol{\omega}_s$, as in Eq. (7.8). So the given angular velocity $\boldsymbol{\omega}$ should be equal to the sum of $\boldsymbol{\omega}_s$ generated by the sliding motion and the angular velocity $\boldsymbol{\omega}_r$ generated by the rolling motion. That is

$$\boldsymbol{\omega} = \boldsymbol{\omega}_s + \boldsymbol{\omega}_r \tag{7.9}$$

We have detailed the procedure to obtain three scalar — the rolling direction φ, rolling rate σ, and compensatory spin rate ω_{spin} — to generate the desired angular velocity by rolling motion in Chapter 6.

Hence, we have solved the inverse kinematics of the rolling–sliding contact. That is, given the translational velocity and angular velocity in Eq. (7.1), we obtain five entities, which are the sliding rate ρ, sliding direction δ, rolling direction φ, rolling rate σ, and compensatory spin rate ω_{spin}, to generate the desired velocity of the surface S'.

7.5 Example of Application: A Unit Sphere Rolling–Sliding on a Paraboloid

Consider a unit sphere rolling–sliding on a paraboloid, as shown in Fig. 7.3.

We obtain the solution to the inverse kinematics when the translational velocity and angular velocity of the sphere are given as

$$\mathbf{v} = v_1\mathbf{e}_u + v_2\mathbf{e}_v$$

$$\boldsymbol{\omega} = \omega_1\mathbf{e}_u + \omega_2\mathbf{e}_v + \omega_3\mathbf{e}_3$$

The rolling rate ρ and rolling direction δ can be obtained straightforwardly as in Eq. (7.3). We now calculate the induced angular velocity $\boldsymbol{\omega}_s$.

The paraboloid can be parameterized as

$$\mathbf{x}(u,v) = \left[u\cos v,\ ,u\sin v,\ \tfrac{u^2}{2}\right]$$

The curvatures of the u- and v-curves have been obtained in Section 6.4.

The normal curvature k_{ns}, geodesic torsion τ_{gs}, and geodesic curvature k_{gs}, in the direction of $\mathbf{e}_{1s}$ can be obtained from Eq. (2.62) as

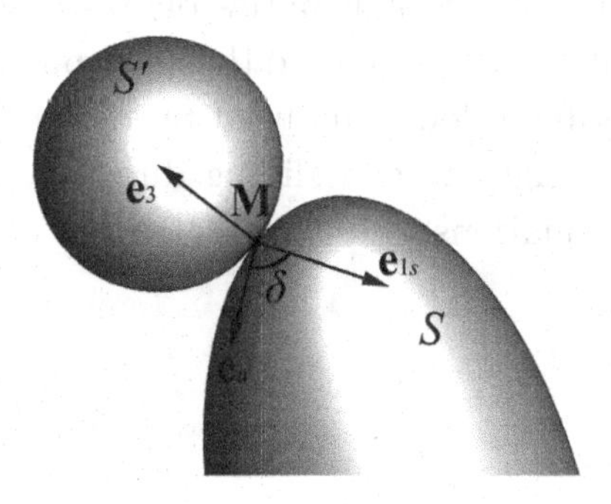

Fig. 7.3: A unit sphere rolling–sliding on a paraboloid

$$\begin{aligned} k_n &= \frac{\cos^2\delta}{\sqrt{(u^2+1)^3}} + \frac{\sin^2\delta}{\sqrt{u^2+1}} \\ k_g &= \frac{\sin\delta}{u\sqrt{u^2+1}} + \frac{d\delta}{ds_s} \\ \tau_g &= 0 \end{aligned} \tag{7.10}$$

The induced angular velocity can be obtained from Eqs. (7.8) and (7.10) as

$$\begin{aligned} \omega_s =& -\rho\left(\frac{\cos^2\delta}{\sqrt{(u^2+1)^3}} + \frac{\sin^2\delta}{\sqrt{u^2+1}}\right)\mathbf{e}_2 \\ &+ \left(\rho\frac{\sin\delta}{u\sqrt{u^2+1}} + \frac{d\delta}{dt}\right)\mathbf{e}_3 \end{aligned} \tag{7.11}$$

where

$$\rho = \sqrt{v_1^2 + v_2^2},\ \delta = \arctan\frac{v_2}{v_1}$$

The rolling motion would need to achieve the difference between the given angular velocity and the induced angular velocity, which is $\omega - \omega_s$. We have demonstrated the approach for this purpose in Chapter 6, thus we skip the steps here.

7.6 Conclusion

We have solved the inverse kinematics of two surfaces maintaining rolling–sliding contact. We have used the fixed-point conditions again to obtain the angular velocity induced by the sliding motion. Two more variables, sliding rate and sliding direction, are then used to solve the contact equations.

Part IV

Kinematics of In-Hand Manipulation

Chapter 8

Kinematic Analysis of the Metahand with Fixed-Point Contact

The Metahand—metamorphic robotic hand—was developed based on the concept of metamorphosis stemming from origami folding at King's College London [Dai *et al.*, 2009; Cui and Dai, 2012; Cui *et al.*, 2017]. Different from other robotic hands, The Metahand has an articulated palm formed by a spherical linkage, enabling the palm to change mobility, topology, and configuration. The Metahand is capable of emulating more sophisticated grasping and dexterous manipulation. It also allows the robotic hand to be fully folded to pass through tight spaces and to be redeployed to adapt for various task requirements.

The Metahand comprises an articulated palm on which three fingers are attached. The base link of the palm is l_1, at P_1 on which the first finger f_1 is mounted. Two input links are on both sides of the base link l_1. The input link l_2 is connected to l_1 via the input joint A on the right side of l_1, and the second finger f_2 is mounted at P_2 of this link. Another input link l_5 is connected to the base link l_1 via the input joint E on the left side of l_1. The third finger f_3 is mounted on the palm link l_3 at P_3, as shown in Fig. 8.1.

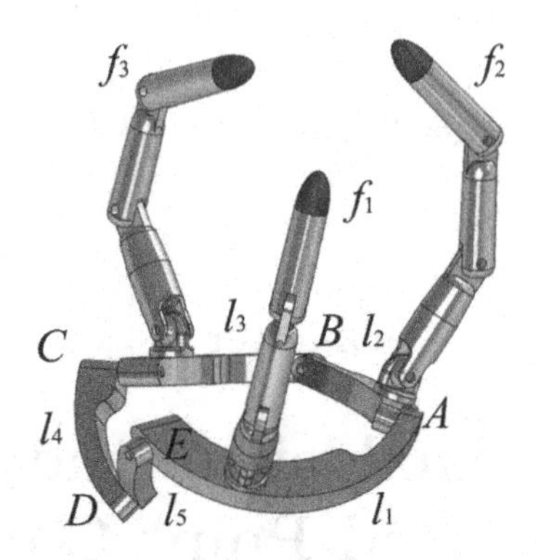

Fig. 8.1: The articulated palm and three fingers of the Metahand

8.1 Problem Definition

We consider the case when the fingertips and the object maintain fixed-point contact, i.e., no relative motion at the contact points. We use hypothetical spherical joints to express the DOFs at the contact points of the object with the three fingertips.

We set up a mathematical model with a finger constraint equation based on screw systems. We use the singular value decomposition (SVD) to obtain the inverse kinematics in closed form and further investigate singularity avoidance by the redundant joints. Finally we obtain the singularity-free range of joint variables.

8.2 Geometric Constraint and Motion Characteristics of the Articulated Palm

We associate screws $\xi_1 - \xi_5$ respectively to the joints $A - E$ revolute joints of the palm, which is based on a five-bar spherical linkage. The joints A and B are active joints and the other three joints C, D, and E are passive joints. The joint angles of $\xi_1 - \xi_5$ are $\theta_1 - \theta_5$, and the central angles of the links $l_1 - l_5$ are $\alpha_1 - \alpha_5$, as shown in Fig. 8.2.

The motion of the palm determines the relative configurations of the three fingers, so we obtain the inverse kinematics of the palm, i.e., obtaining θ_1 in terms of θ_2 and θ_3. For this purpose, we set up

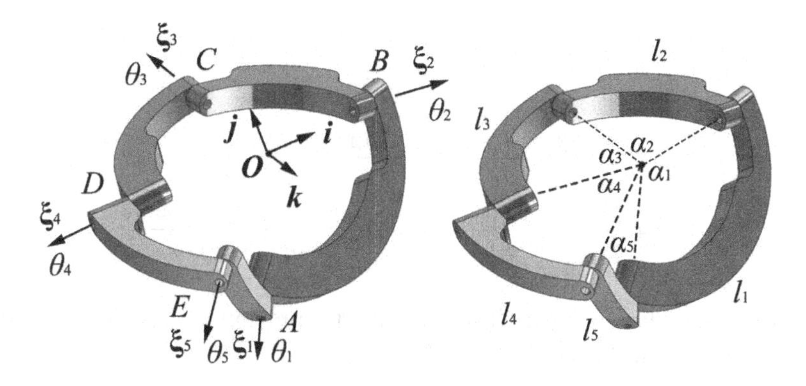

Fig. 8.2: The screw systems of the palm

a frame (**O**, **i**, **j**, **k**) fixed on the base link l_1 in such a way that **O** represents the center of the palm linkage, the **i**-axis is along OA, the **k**-axis is perpendicular to the OAB plane, which is the base link plane, as shown in Fig. 8.2.

Given the value of the input angles θ_2 and θ_3, the coordinates $(\mathbf{x}_D, \mathbf{y}_D, \mathbf{z}_D)$ of the point D can be readily obtained as

$$
\begin{aligned}
x_D &= \cos\alpha_2\cos\alpha_3 - \sin\alpha_2\sin\alpha_3\cos\theta_3 \\
y_D &= \sin\alpha_2\cos\alpha_3\cos\theta_2 \\
&\quad + (\cos\alpha_2\cos\theta_2\cos\theta_3 - \sin\theta_2\sin\theta_3)\sin\alpha_3 \\
z_D &= \sin\alpha_2\cos\alpha_3\sin\theta_2 \\
&\quad + (\cos\alpha_2\sin\theta_2\cos\theta_3 + \cos\theta_2\sin\theta_3)\sin\alpha_3
\end{aligned}
\tag{8.1}
$$

Let (x_E, y_E, z_E) and (x_A, y_A, z_A) represent the coordinates of the points E and A, respectively. The geometric constraints of the spherical palm give the following constraint equations:

$$
\begin{aligned}
x_D x_E + y_D y_E + z_D z_E &= \cos\alpha_4 \\
x_E x_A + y_E y_A + z_E z_A &= \cos\alpha_5 \\
x_E^2 + y_E^2 + z_E^2 &= 1
\end{aligned}
\tag{8.2}
$$

The first two equations in Eq. (8.2) yield the expressions of x_E and y_E in terms of z_E as

$$\begin{aligned} x_E &= \frac{z_D y_A - y_D z_A}{x_D y_A - y_D x_A} z_E + \frac{y_A \cos\alpha_4 - y_D \cos\alpha_5}{x_D y_A - y_D x_A} \\ y_E &= \frac{x_A z_D - z_A x_D}{x_D y_A - y_D x_A} z_E + \frac{x_D \cos\alpha_5 - x_A \cos\alpha_4}{x_D y_A - y_D x_A} \end{aligned} \tag{8.3}$$

Substituting Eq. (8.3) into the third equation of Eq. (8.2) yields a quadratic equation in terms of z_D:

$$A_1 z_E^2 + A_2 z_E + A_3 = 0 \tag{8.4}$$

where

$$\begin{aligned} A_1 &= \frac{(z_D y_A - y_D z_A)^2}{(x_D y_A - y_D x_A)^2} + \frac{(x_A z_D - z_A x_D)^2}{(x_D y_A - y_D x_A)^2} + 1 \\ A_2 &= \frac{2(y_D \cos\alpha_5 - y_A \cos\alpha_4)(z_D y_A - y_D z_A)}{(x_D y_A - y_D x_A)^2} + \\ &\quad \frac{2(x_D \cos\alpha_5 - x_A \cos\alpha_4)(z_D x_A - x_D z_A)}{(x_D y_A - y_D x_A)^2} \\ A_3 &= \frac{(y_D \cos\alpha_5 - y_A \cos\alpha_4)^2}{(x_D y_A - y_D x_A)^2} + \frac{(x_D \cos\alpha_5 - x_A \cos\alpha_4)^2}{(x_D y_A - y_D x_A)^2} - 1 \end{aligned}$$

Two solutions exist for Eq. (8.4), which corresponds to two assembly configurations of the five-bar palm. However, only one solution can be chosen as it is impossible to change from one assembly configuration to the other without disassembling and reassembling the palm linkage. The mechanical structure of the palm determines the coordinate z_E as

$$z_D = \frac{-A_2 + \sqrt{A_2^2 - 4A_1 A_2}}{2A_1} \tag{8.5}$$

The coordinates x_E and y_E can be obtained from Eq. (8.3) straightforwardly, where detailed calculations are omitted.

On the other hand, the coordinates (x_E, y_E, z_E) can be expressed in terms of θ_1 as

$$\begin{bmatrix} x_E \\ y_E \\ z_E \end{bmatrix} = \begin{bmatrix} -\sin\alpha_1 \cos\alpha_5 - \cos\alpha_1 \sin\alpha_5 \cos\theta_1 \\ -\cos\alpha_1 \cos\alpha_5 + \sin\alpha_1 \sin\alpha_5 \cos\theta_1 \\ -\sin\alpha_5 \sin\theta_1 \end{bmatrix} \tag{8.6}$$

The passive palm angle θ_1 can be obtained from Eqs. (8.5) and (8.6) as

$$\theta_1 = \arcsin \frac{A_2 - \sqrt{A_2^2 - 4A_1A_3}}{2A_1 \sin\alpha_5} \tag{8.7}$$

The other two joint angles θ_4 and θ_5 can be readily obtained by using the geometric constraints of the link lengths.

Note that any two joint angles can be used as inputs to obtain the other three joint angles. This completes the kinematics of the articulated palm.

8.3 The Kinematic Characteristic Equation of the Metamorphic Hand with Fixed-Point Contact

We assume that the fingertips and the object maintain fixed-point contact and the contact points are within the workspace of the Metahand. We further use hypothetical joint to present the relative DOFs at the contact of the object with each fingertip.

Examining the mechanical structure of the Metahand shows that the finger f_1 is symmetric to the finger f_3 with respect to the finger f_2. Let ξ_{obj} represent the instantaneous motion of the object and ξ_{ij} the screw of the jth joint of the ith finger, as shown in Fig. 8.3.

This twist can be expressed as a linear combination of the joint screws and the screws representing the DOFs of the hypothetical spherical joints of the fingers. For the finger f_1, the following relationship exists:

$$\begin{aligned} \xi_{\text{obj}} &= \dot{\theta}_2\xi_2 + \dot{\theta}_{1,1}\xi_{1,1} + \dot{\theta}_{1,2}\xi_{12} + \dot{\theta}_{1,3}\xi_{13} \\ &\quad + (\dot{\theta}_{1,4}\xi_{14} + \dot{\theta}_{1,5}\xi_{15} + \dot{\theta}_{1,6}\xi_{16}) \end{aligned} \tag{8.8}$$

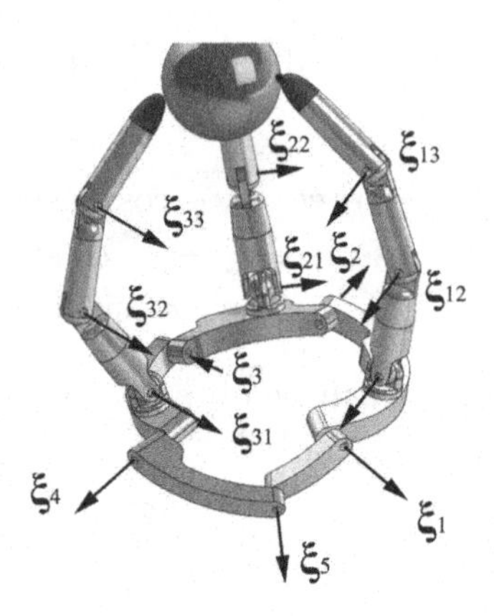

Fig. 8.3: Screws of the Metahand joints

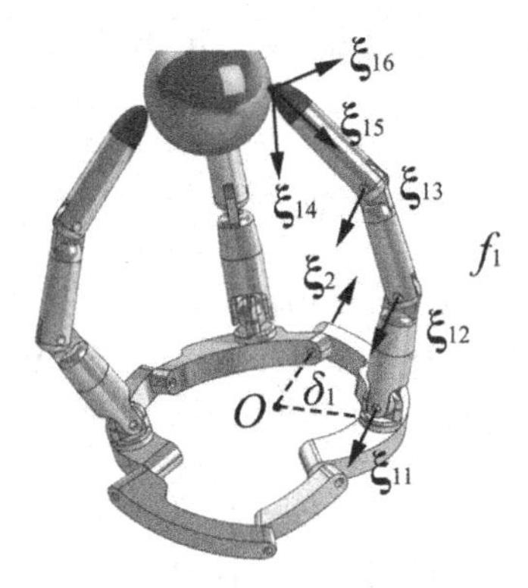

Fig. 8.4: The screw system of the finger f_1

where ξ_{14} – ξ_{16} are the screws representing the DOFs of the hypothetical spherical joints at the contact point between the finger f_1 and the object, and $\dot{\theta}_{1,1}$ to $\dot{\theta}_{1,6}$ represent the joint velocities of the hypothetical spherical joint, as shown in Fig. 8.4.

Similarly, the twist of the object is related to the screws of the fingers f_2 and f_3 as

$$\begin{aligned}\xi_{\text{obj}} &= \dot{\theta}_{2,1}\xi_{21} + \dot{\theta}_{2,2}\xi_{22} + \dot{\theta}_{2,3}\xi_{23} \\ &\quad + (\dot{\theta}_{2,4}\xi_{24} + \dot{\theta}_{2,5}\xi_{25} + \dot{\theta}_{2,6}\xi_{26})\end{aligned}$$

$$\xi_{\text{obj}} = \dot{\theta}_3\xi_3 + \dot{\theta}_{3,1}\xi_{31} + \dot{\theta}_{3,2}\xi_{32} + \dot{\theta}_{3,3}\xi_{33} + (\dot{\theta}_{3,4}\xi_{34} + \dot{\theta}_{3,5}\xi_{35} + \dot{\theta}_{3,6}\xi_{36}) \tag{8.9}$$

Let $\xi^r_{i1} - \xi^r_{i3}$ denote the reciprocal screws of the hypothetical spherical joints $\xi_{i4} - \xi_{i6}$. Taking the reciprocal product of both sides of Eq. (8.8) with each reciprocal screw $\xi^r_{11} - \xi^r_{13}$ yield three equations, eliminating the terms related to the hypothetical spherical joints. We use the same procedure to the two equations in Eq. (8.9). In this way, we obtain nine equations in matrix form as

$$\mathbf{J}_x^T \Delta\xi_{\text{obj}} = \mathbf{J}_q \dot{\theta} \tag{8.10}$$

where

$$\mathbf{J}_x^T = \begin{bmatrix} \xi^{r\,T}_{11} \\ \xi^{r\,T}_{12} \\ \vdots \\ \xi^{r\,T}_{33} \end{bmatrix}, \quad \mathbf{J}_q = \begin{bmatrix} \mathbf{J}_q^1 & 0 & 0 \\ 0 & \mathbf{J}_q^2 & 0 \\ 0 & 0 & \mathbf{J}_q^3 \end{bmatrix}, \quad \dot{\theta} = \begin{bmatrix} \dot{\theta}_2 \\ \dot{\theta}_{1,1} \\ \vdots \\ \dot{\theta}_{3,3} \end{bmatrix}$$

This gives the characteristic kinematic equation of the Metahand, which describes its instantaneous kinematics. The Jacobian matrices $\mathbf{J}_x^T$ and $\mathbf{J}_q$ of the Metahand is hence given based on the reciprocity of screw systems. The matrix $\mathbf{J}_x^T$ gives the forward Jacobian matrix and $\mathbf{J}_q$ gives the inverse Jacobian matrix induced. On the left-hand side of Eq. (8.10) $\mathbf{J}_x^T$ is a 9×6 matrix, and on the right-hand side $\mathbf{J}_q$ is a 9×11 matrix.

The forward kinematics is to obtain the object twist ξ_{obj} given the joint velocities $\dot{\theta}_2$ to $\dot{\theta}_{3,3}$. In this case, Eq. (8.10) becomes an overdetermined linear system of equations in which the number of equations is more than the number of variables. The solutions can only be found approximately: usually the residue is minimized in the context of least squares.

The inverse kinematics amounts to obtaining the joint velocities $\dot{\theta}_2$ to $\dot{\theta}_{3,3}$ given the object twist ξ_{obj}. In this case, Eq. (8.10) becomes an under-determined linear system of equations, which has fewer equations than the number of variables. Thus, an infinite number of solutions exist.

8.4 Reciprocity-Based Jacobian Matrix and the Finger Constraint Equation

8.4.1 *Constraint Equations of the Fingers*

Based on the reciprocity analysis, each sub-matrix $\mathbf{J}_q^i$ of the ith finger can be obtained as

$$\begin{aligned}
\mathbf{J}_q^1 &= \begin{bmatrix} \xi_{11}^{r\ T} \\ \xi_{12}^{r\ T} \\ \xi_{13}^{r\ T} \end{bmatrix} \Delta[\xi_2 \quad \xi_{11} \quad \xi_{12} \quad \xi_{13}] \\
\mathbf{J}_q^2 &= \begin{bmatrix} \xi_{21}^{r\ T} \\ \xi_{22}^{r\ T} \\ \xi_{23}^{r\ T} \end{bmatrix} \Delta[\xi_{21} \quad \xi_{22} \quad \xi_{23}] \\
\mathbf{J}_q^3 &= \begin{bmatrix} \xi_{31}^{r\ T} \\ \xi_{32}^{r\ T} \\ \xi_{33}^{r\ T} \end{bmatrix} \Delta[\xi_3 \quad \xi_{31} \quad \xi_{32} \quad \xi_{33}]
\end{aligned} \tag{8.11}$$

It can be seen that the inverse kinematics of each finger is independent of the others. The inverse kinematics of the finger f_1 is similar to that of the finger f_3 and the solution to inverse kinematics of the finger f_1 can be applied to that of the finger f_3. We thus derive the kinematics of the finger f_1 as a case study.

Selecting the reference system carefully could greatly simplify the derivation of the screw system and the inverse kinematics in velocity domain. Thus, a local reference system $(\mathbf{P}_1 - \mathbf{i}_1\mathbf{j}_1\mathbf{k}_1)$ is defined with the $\mathbf{i}_1$-axis along OP_1 and $\mathbf{j}_1$-axis perpendicular to AOE plane of the link l_1. The angle between OP_1 and OA is δ_1, as shown in Fig. 8.5.

The screws ξ_2 and ξ_{11} – ξ_{13} can be obtained w.r.t. the frame $(\mathbf{P}_1 - \mathbf{i}_1\mathbf{j}_1\mathbf{k}_1)$ as

$$\xi_2 = \begin{bmatrix} \cos\delta_1 \\ 0 \\ -\sin\delta_1 \\ 0 \\ -R\sin\delta_1 \\ 0 \end{bmatrix}, \quad \xi_{11} = \begin{bmatrix} 0 \\ 0 \\ 1 \\ 0 \\ 0 \\ 0 \end{bmatrix},$$

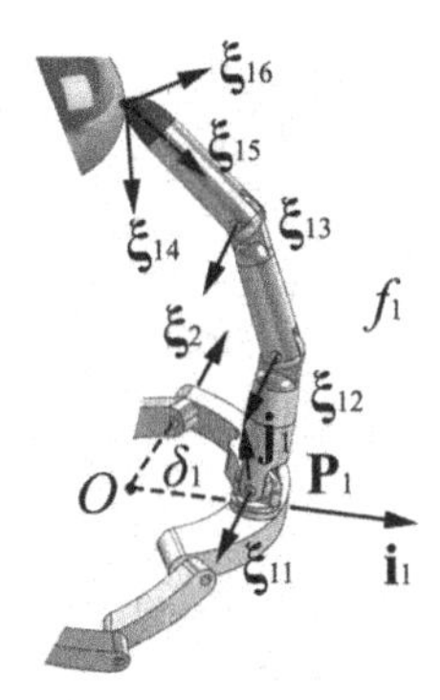

Fig. 8.5: A local reference frame of the finger f_1

$$\xi_{12} = \begin{bmatrix} 0 \\ 0 \\ 1 \\ a_1 \sin\theta_{1,12} \\ -a_1 \cos\theta_{1,12} \\ 0 \end{bmatrix}, \quad \xi_{13} = \begin{bmatrix} 0 \\ 0 \\ 1 \\ a_1 \sin\theta_{11} + a_2 \sin\theta_{1,12} \\ -a_1 \cos\theta_{11} - a_2 \cos\theta_{1,12} \\ 0 \end{bmatrix} \tag{8.12}$$

where R is the radius of the spherical palm, $a_1 - a_3$ are the lengths of three finger links respectively, and the $\theta_{1,12} = \theta_{1,1} + \theta_{1,2}$.

A hypothetical spherical joint can be represented as three revolute joints meeting at the contact point between the object and the fingertip. The three screws of a hypothetical spherical joint are self-reciprocal. Reciprocal screws of a hypothetical spherical joint can take any direction as long as they pass through the contact point and are linearly independent.

We aim to obtain a symbolic solution to the inverse kinematics, so we choose the reciprocal screws of the hypothetical spherical joint as

$$\xi_{11}^r = \begin{bmatrix} \cos\theta_{1,123} \\ \sin\theta_{1,123} \\ 0 \\ 0 \\ 0 \\ a_1 \sin\theta_{1,123} + a_2 \sin\theta_{1,123} \end{bmatrix}$$

$$\xi_{12}^r = \begin{bmatrix} -\sin\theta_{1,123} \\ \cos\theta_{1,123} \\ 0 \\ 0 \\ 0 \\ a_1\cos\theta_{1,123} + a_2\cos\theta_{1,3} + a_3 \end{bmatrix}$$

$$\xi_{13}^r = \begin{bmatrix} 0 \\ 0 \\ 1 \\ a_1\sin\theta_{1,1} + a_2\sin\theta_{1,12} + a_3\sin\theta_{1,123} \\ -(a_1\cos\theta_{1,1} + a_2\cos\theta_{1,12} + a_3\cos\theta_{1,123}) \\ 0 \end{bmatrix} \tag{8.13}$$

$\theta_{1,123} = \theta_{1,1} + \theta_{1,2} + \theta_{1,3}$. The three screws in Eq. (8.13) pass through the contact point.

Taking the reciprocal product on both sides of Eq. (8.8) by the three reciprocal screws ξ_{11}^r, ξ_{12}^r, ξ_{13}^r yields the constraint equation of the finger f_1 as

$$\begin{bmatrix} m_{11} & 0 & 0 & 0 \\ m_{21} & m_{22} & m_{23} & 0 \\ m_{31} & m_{32} & m_{33} & m_{34} \end{bmatrix} \begin{bmatrix} \dot\theta_2 \\ \dot\theta_{1,1} \\ \dot\theta_{1,2} \\ \dot\theta_{1,3} \end{bmatrix} = \begin{bmatrix} g_{11} \\ g_{12} \\ g_{13} \end{bmatrix} \tag{8.14}$$

where

$$\begin{aligned}
m_{11} &= \cos\delta_1(a_1\sin\theta_{1,1} + a_2\sin\theta_{1,12} + a_3\sin\theta_{1,123}) \\
m_{21} &= -\sin\delta_1(a_1\sin\theta_{1,23} + a_2\sin\theta_{1,3} + R\sin\theta_{1,123}) \\
m_{22} &= a_1\sin\theta_{1,23} + a_2\sin\theta_{1,3} \\
m_{23} &= a_2\sin\theta_{1,3} \\
m_{31} &= -\sin\delta_1(a_1\cos\theta_{1,23} + a_2\cos\theta_{1,3} + R\cos\theta_{1,123}) \\
m_{32} &= a_1\cos\theta_{1,23} + a_2\cos\theta_{1,3} + a_3 \\
m_{33} &= a_2\cos\theta_{1,3} + a_3 \\
m_{34} &= a_3
\end{aligned}$$

The elements $g_{11} - g_{13}$ on the right-hand side of Eq. (8.14) are equal to the values of the reciprocal product of $\xi_{11}^r - \xi_{13}^r$ and the twist of the object w.r.t. the frame $(\mathbf{P}_1 - \mathbf{i}_1\mathbf{j}_1\mathbf{k}_1)$.

8.4.2 *Finger-Joint Velocities Based on SVD*

The reciprocity analysis generates the finger constraint matrix in Eq. (8.14), which can be partitioned into the sub-matrices as

$$\begin{bmatrix} m_{11} & 0 & 0 & 0 \\ m_{21} & m_{22} & m_{23} & 0 \\ m_{31} & m_{32} & m_{33} & m_{34} \end{bmatrix} = \begin{bmatrix} \mathbf{J}_{11} & 0 \\ \mathbf{J}_{21} & \mathbf{J}_{22} \end{bmatrix} \tag{8.15}$$

where

$$\mathbf{J}_{11} = \begin{bmatrix} m_{11} & 0 & 0 \\ m_{21} & m_{22} & m_{23} \end{bmatrix}, \quad \mathbf{J}_{21} = \begin{bmatrix} m_{31} & m_{32} & m_{33} \end{bmatrix}, \quad \mathbf{J}_{22} = \begin{bmatrix} m_{34} \end{bmatrix}$$

The sub-matrix $\mathbf{J}_{11}$ can be further partitioned into

$$\mathbf{J}_{11} = \begin{bmatrix} \mathbf{K}_{11} & 0 \\ \mathbf{K}_{21} & \mathbf{K}_{22} \end{bmatrix} \tag{8.16}$$

where

$$\mathbf{K}_{11} = \begin{bmatrix} m_{11} \end{bmatrix}, \quad \mathbf{K}_{21} = \begin{bmatrix} m_{21} \end{bmatrix}, \quad \mathbf{K}_{22} = \begin{bmatrix} m_{22} & m_{23} \end{bmatrix}$$

Hence, Eq. (8.14) can be rewritten as

$$\begin{aligned} \mathbf{K}_{11}\dot{\theta}_2 &= g_{11} \\ \mathbf{K}_{22}\begin{bmatrix} \dot{\theta}_{1,1} \\ \dot{\theta}_{1,2} \end{bmatrix} &= g_{12} - \mathbf{K}_{21}\dot{\theta}_2 \\ \mathbf{J}_{22}\dot{\theta}_{2,3} &= g_{13} - \mathbf{J}_{21}\begin{bmatrix} \dot{\theta}_2 \\ \dot{\theta}_{1,1} \\ \dot{\theta}_{1,2} \end{bmatrix} \end{aligned} \tag{8.17}$$

The singular values σ_{11}, σ_{12}, and σ_{13} of the Jacobian matrix of the finger f_1 can be obtained from its sub-matrices $\mathbf{K}_{11}$, $\mathbf{K}_{22}$, and

$\mathbf{J}_{22}$ of Eq. (8.17) as

$$\begin{aligned}
\sigma_{11} &= m_{11} = \cos\delta_1(a_1 \sin\theta_{1,1} + a_2 \sin\theta_{1,12} + a_3 \sin\theta_{1,123}) \\
\sigma_{12} &= \sqrt{m_{22}^2 + m_{23}^2} = \sqrt{(a_1 \sin\theta_{1,23} + a_2 \sin\theta_{1,3})^2 + (a_2 \sin\theta_{1,3})^2} \\
\sigma_{13} &= m_{34} = a_3
\end{aligned} \tag{8.18}$$

Equation (8.18) is an under-determined system of linear equations. If we set the joint $\theta_{1,2}$ as a redundant joint, we can obtain the joint velocities of the finger f_1 in terms of these singular values as

$$\begin{aligned}
\dot{\theta}_2 &= \frac{g_{11}}{\sigma_1} \\
\dot{\theta}_{1,1} &= \frac{g_{12} - m_{21}\dot{\theta}_2}{\sigma_2} \\
\dot{\theta}_{1,3} &= \frac{1}{a_3}\left(g_{13} - \begin{bmatrix} m_{31} & m_{32} & m_{33} \end{bmatrix} \begin{bmatrix} \dot{\theta}_2 \\ \dot{\theta}_{1,1} \\ \dot{\theta}_{1,2} \end{bmatrix}\right)
\end{aligned} \tag{8.19}$$

8.5 Singularity Avoidance by Redundancy

The finger joint velocities are expressed in terms of the singular values. They are functions of the finger joint angles $\theta_{1,1}$, $\theta_{1,2}$, and $\theta_{1,3}$, each of which moves in the following ranges respectively:

$$\theta_{1,1} \in [\pi/4, 3\pi/4], \quad \theta_{1,2} \in [-\pi/4, 3\pi/4], \quad \theta_{1,3} \in [\pi/4, 3\pi/4]$$

Singularities occur when any of the singular values σ_{11} – σ_{13} is 0. The singular value σ_{13} is a non-zero constant. The singular value σ_{12} is 0 only when both the joint angles $\theta_{1,2}$ and $\theta_{1,3}$ are simultaneously 0, which can be deduced straightforwardly from the second equation in Eq. (8.18). Thus, we can make the redundant joint $\theta_{1,2}$ nonzero to avoid this singularity.

We now check the singular value σ_{11}. The gradient of σ_{11} can be obtained as

$$\nabla\sigma_{11} = \begin{bmatrix} \cos\delta_1(a_1\cos\theta_{1,1} + a_2\cos\theta_{1,12} + a_3\cos\theta_{1,123}) \\ \cos\delta_1 a_3\cos\theta_{1,123} \end{bmatrix} \tag{8.20}$$

The Hessian matrix of σ_{11} can be obtained as

$$H(\sigma_{11}) = \begin{bmatrix} -\sigma_{11} & -a_3\cos\delta_1\sin\theta_{1,123} \\ -a_3\cos\delta_1\sin\theta_{1,123} & -a_3\cos\delta_1\sin\theta_{1,123} \end{bmatrix} \tag{8.21}$$

Equating the gradient in Eq. (8.20) to 0 yields two solutions:

$$\theta_{1,123} = \frac{\pi}{2}, \quad \theta_{1,1} = \frac{\pi - \theta_{1,2}}{2}$$

$$\theta_{1,123} = \frac{3\pi}{2}, \quad \theta_{1,1} = \frac{\pi - \theta_{1,2}}{2}$$

Both the determinants of the Hessian matrix in Eq. (8.21) in the two cases are negative, implying that the singular value σ_{11} is concave in terms of $\theta_{1,1}$ and $\theta_{1,3}$.

We can use the values of the redundant joint angle $\theta_{1,2}$ to shift the values of σ_{11} above 0 for singularity avoidance. We have found that there is no singularity when $\theta_{1,2}$ is in the range of $[-\pi/4, 1.86]$, which can be verified by Figs. 8.6(a)–8.6(c).

Figure 8.6(a) shows the singular value σ_{11} does not have 0 value when $\theta_{1,2} = 0.52$ and both $\theta_{1,1}$ and $\theta_{1,3}$ in the range of $[-\pi/4, 3\pi/4]$. When the values of $\theta_{1,2}$ increase to half π, the singular value σ_{11} starts to have 0 value, as shown in Fig. 8.6(b). Figure 8.6(c) shows σ_{11} has 0 only when $\theta_{1,2} = 2.618$. Hence, the singularities of the finger f_1 can be avoided if $\theta_{1,2}$ is confined in the ranges of $\pi/4 - 0$ and $0 - 1.86$.

The non-singular range of the fingers f_2 and f_3 can be obtained in a similar manner. In this way, the inverse kinematics of the Metahand does not have any singularities by using the redundant joints $\theta_{1,2}$, $\theta_{2,2}$, and $\theta_{3,2}$.

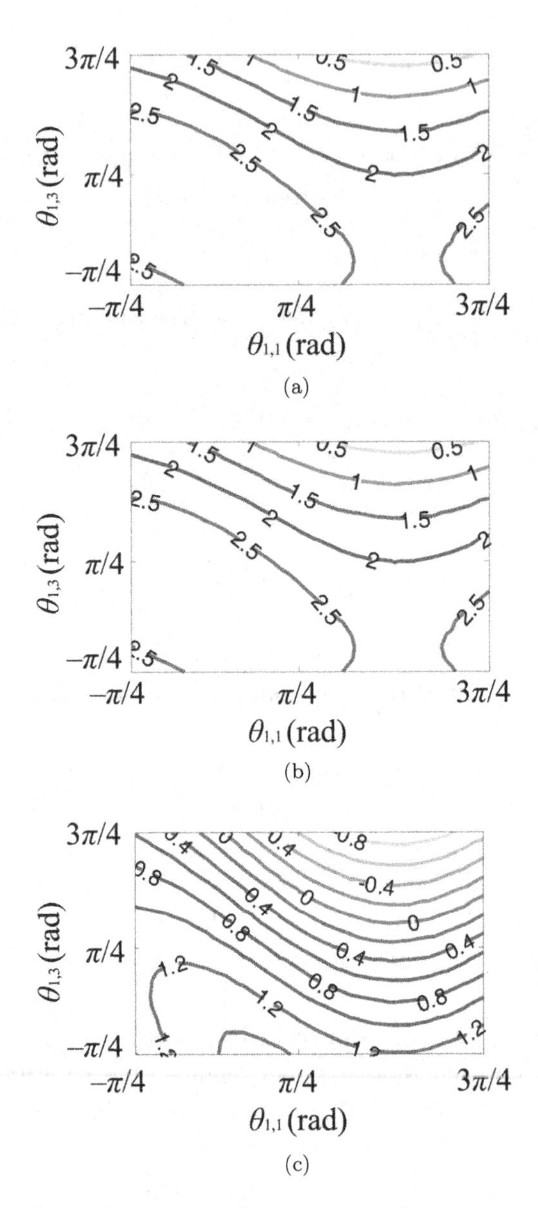

Fig. 8.6: The values of singular value σ_{11} vary when the joint angle $\theta_{1,2}$ in different configurations: (a) Level set of singular value σ_{11} in the case of $\theta_{1,2} = 0.52$; (b) Level set of singular value σ_{11} in the case of $\theta_{1,2} = \pi/2$; (c) Level set of singular value σ_{11} in the case of $\theta_{1,2} = 3\pi/2$.

8.6 Conclusion

We have obtained the kinematics of in-hand manipulation of the Metahand when the fingertips and the object maintain fixed-point contact. We have established a framework for the kinematics of in-hand manipulation when the fingertips and the object maintain the fixed-point contact. This framework facilitates singularity avoidance via SVD.

Chapter 9

Workspace and Posture Analysis of the Metahand

In this chapter, we investigate the workspace of the articulated palm and the posture workspace of the Metahand. We apply the concept of Gauss map to the finger-operation plane and visualize the triangular palm workspace and helical posture workspace.

9.1 Problem Definition

We analyze the workspace and posture of the Metahand to understand the space that the fingers can reach and the range of postures that the Metahand can create with the articulated palm.

We introduce the concept of finger-operation plane relating the finger motion to the palm motion and the postures of the hand. Gauss map from differential geometry is used to represent the changes of these planes when palm inputs vary.

We further introduce the concept of palm workspace triangle, each vertex of which corresponds to the location of each finger is mounted on the palm. The palm workspace triangle evolves into a helical triangular tube when palm inputs vary.

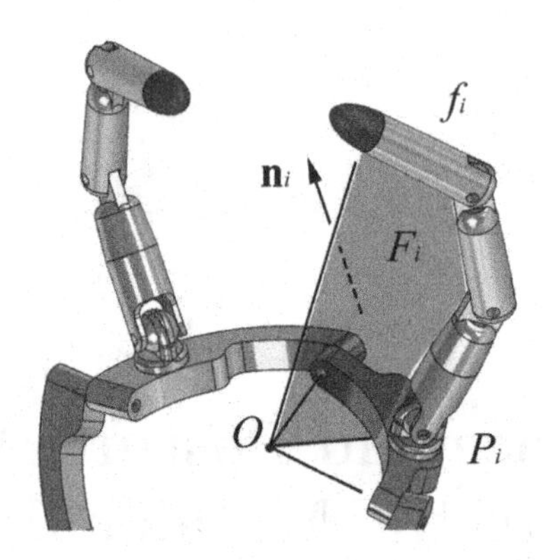

Fig. 9.1: The finger-operation plane of the finger f_i

9.2 The Finger-Operation Plane

Each finger consists of three parallel revolute joints and works on a plane F, which is formed by finger links and passes the center O of the palm, as shown in Fig. 9.1.

We name this plane the finger-operation plane, which varies with movement of the palm. Each link of the palm together with the center point O of the palm defines a palm-link plane. The finger-operation plane of each finger is always perpendicular to its the palm-link plane where the finger is mounted.

The normals $\mathbf{n}_i$ of the finger-operation planes F_i are determined only by the palm linkage, i.e., they are functions of the two palm input angles θ_1 and θ_2, which are associated with the joints A and B, respectively.

9.3 Posture Mapping and the Posture Ruled Surfaces

A set of normals $\mathbf{n}_1$, $\mathbf{n}_2$, and $\mathbf{n}_3$ can be obtained when the values of input angles θ_1 and θ_2 are specified. To visually present a configuration of the Metahand posture, we use the concept of Gauss map to map these three normals onto a unit sphere, revealing the lateral relationship among the three finger-operation planes. A particular posture with $\theta_1 = \pi/2$ and $\theta_2 = 0.7$ is shown in Fig. 9.2.

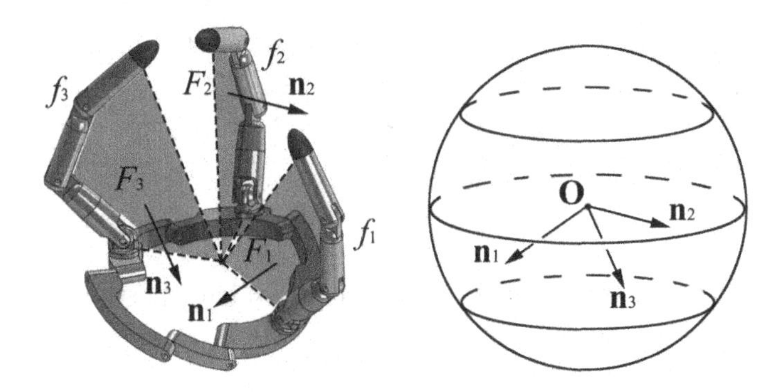

Fig. 9.2: Gauss map of a hand posture

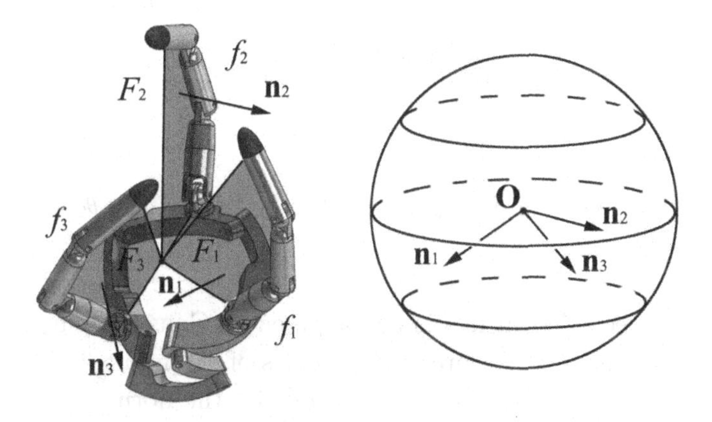

Fig. 9.3: Gauss map of another hand posture

The articulated palm enables the change hand postures, i.e., the relative configurations of the finger-operation planes. For example, if the angle θ_1 of the palm changes from 0 to $5\pi/6$, the hand posture changes accordingly. This new hand posture is represented in another Gauss map, as shown in Fig. 9.3.

The two Gauss maps in Figs. 9.2 and 9.3 can be superimposed along an axis representing the values of θ_1, yielding an augmented Gauss map, as shown in Fig. 9.4.

When the input angle θ_1 varies in its range of motion $[0, 2\pi]$, the normal vectors $\mathbf{n}_1$, $\mathbf{n}_2$, and $\mathbf{n}_3$ of the three finger-operation planes

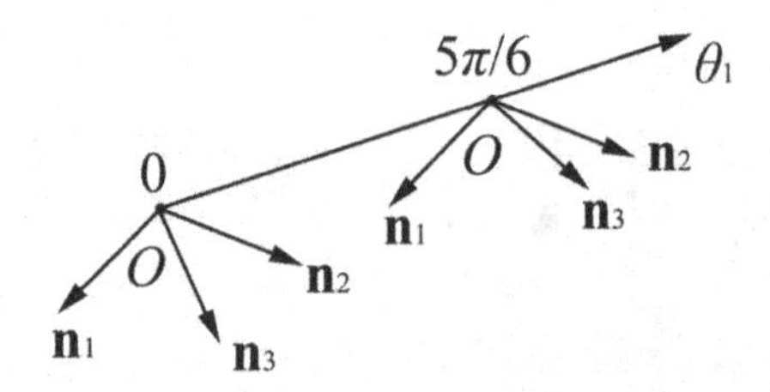

Fig. 9.4: Augmented Gauss map of two hand postures

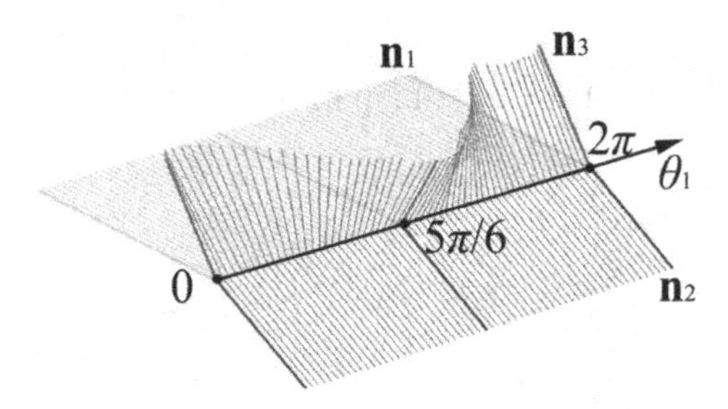

Fig. 9.5: Posture ruled surfaces when $\theta_1 \in [0, 2\pi]$ and $\theta_2 = 0.7$

generate three ruled surfaces in an augmented Gauss map in Fig. 9.5, where the central axis representing values of the input θ_1.

In Fig. 9.5, two planes are swept out by the normals $\mathbf{n}_1$ and $\mathbf{n}_2$ moving along the θ_1-axis. The normal $\mathbf{n}_3$ generates a ruled surface in a similar manner. Two previous hand postures in Figs. 9.2 and 9.3 are marked with thick lines when $\theta_1 = 0$ and $\theta_1 = 5\pi/6$ while θ_2 is fixed at 0.7. These augmented posture ruled surfaces explicitly demonstrate the relative configurations among the finger-operation planes and provide intuitive clues to adjusting hand postures for grasping and manipulation.

When the second input angle θ_2 of the palm varies in the range of $-\pi/2$–$\pi/2$, a set of posture ruled surfaces can be yielded, which is of cylindrical shape. To better illustrate these posture ruled surfaces, we fix the values of θ_2 at $-\pi/2$, 0, and $\pi/2$, and sketch out the three posture ruled surfaces respectively, while the first variable θ_1 of the palm varies from 0 to 2π, as shown in Fig. 9.6.

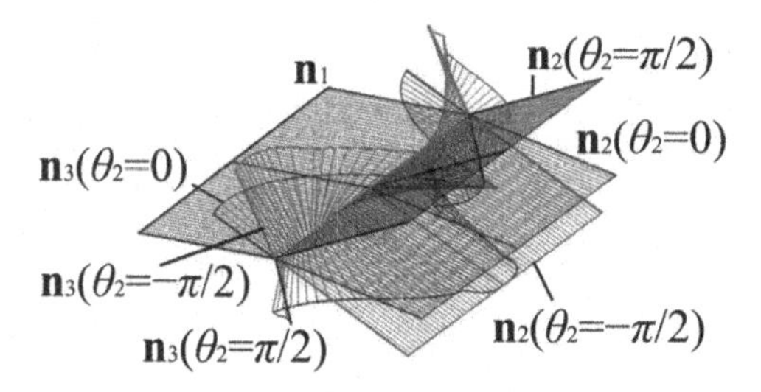

Fig. 9.6: Posture map of the Metahand

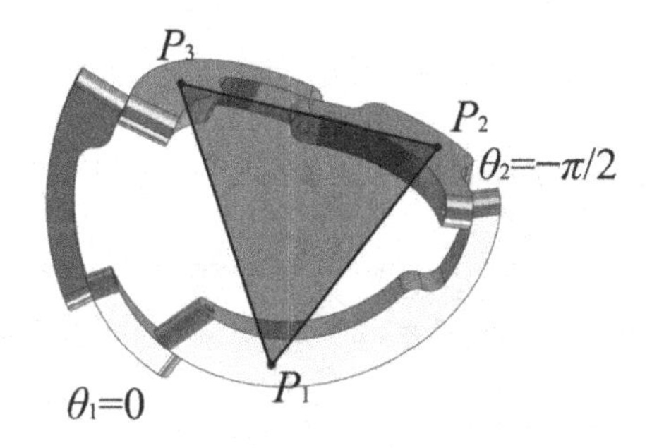

Fig. 9.7: A workspace triangle of the palm when $\theta_1 = 0$ and $\theta_2 = -\pi/2$

9.4 The Workspace of the Metahand Palm

The motion of the Metahand palm affects the workspace of the fingers f_2 and f_3. We define the palm workspace triangle as represented by the triangle determined by the three finger-mounting points P_1, P_2, and P_3, as shown in Fig. 9.7. Note that the palm-workspace triangle varies with the palm motion.

A helical workspace triangle tube can be generated in a similar approach to generating the posture ruled surfaces. We fix the value of θ_2 and set the θ_1-axis with the range of 0–2π, and then map the palm-workspace triangle in such a way that the point P_1 moves on the θ_1-axis with the increase of the value of θ_1, and the triangles $P_1P_2P_3$ are shifted to this space with their relative orientation maintained, as shown in Fig. 9.8.

When the other palm input θ_2 varies from $-pi/2$ to $pi/2$, the palm has 2 DOFs to change the workspace triangle. Fig. 9.9 shows three

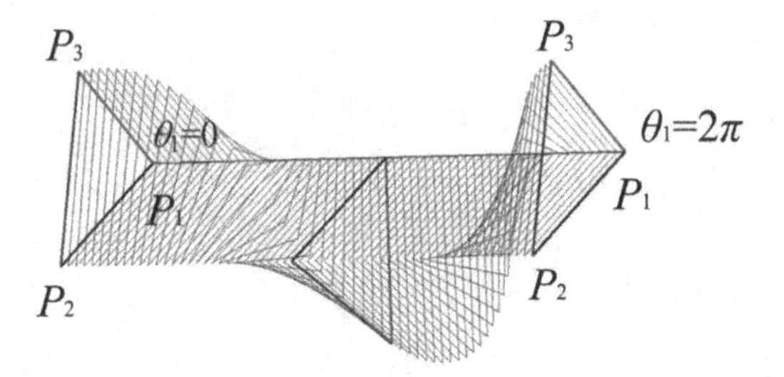

Fig. 9.8: A helical workspace triangular tube when $\theta_1 \in [0, 2\pi]$ and $\theta_2 = -\pi/2$

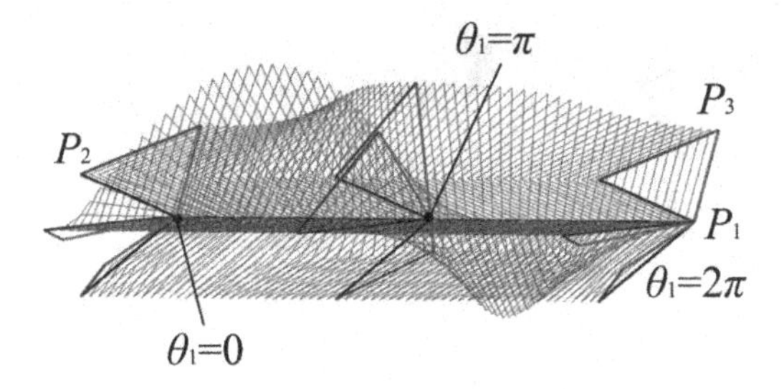

Fig. 9.9: Palm workspace map when $\theta_2 = -\pi/2$, $\theta_2 = 0$, and $\theta_2 = \pi/2$

sets of palm workspace triangle tubes as that in Fig. 9.8 are generated, when θ_2 is set at $-pi/2$, 0, and $pi/2$ while the input variable θ_1 varies from 0 to 2π.

9.5 Conclusion

We have applied the concept of the Gauss map to the finger-operation plane to map the palm postures into posture ruled surfaces and the palm triangular workspace into the palm helical tubes. The posture ruled surfaces and palm helical tubes serve as visual aid for grasp and manipulation tasks.

Chapter 10

Rolling Contact in Kinematics of In-Hand Manipulation

In this chapter, we incorporate the kinematic formulations of rolling contact and in-hand manipulation with fixed-point contact developed in previous chapters into that of in-hand manipulation with rolling contact. The proposed formulation admits a high degree of coordinate independence.

10.1 Problem Definition

We formulate a framework of the forward and inverse velocity kinematics of in-hand manipulation using the product-of-exponentials (POE) formula. The velocity formulation of rolling contact in terms of the Darboux frame facilitates the analysis in the integrated framework. We treat each finger of a robotic hand as an open-chain linkage connected by n revolute joints. When the object surface and the fingertip surface roll on each other, the point of contact on each surface traces a trajectory. Each fingertip must follow the object, maintain contact, and be independent of the other fingers. It therefore suffices to examine the motion of a single finger.

10.2 Forward Velocity Kinematics of In-Hand Manipulation with Rolling Contact

The paths of the contact point across each surface are determined by the geometry of each surface, by the motion of the object, and by the kinematics of the finger. A contact trajectory exists on each of the surfaces, and these contact trajectories and finger joints together determine the motion of the object.

Let P represent a frame fixed on the palm, T a tool frame fixed on the fingertip, B a frame fixed on a point B of the object, ξ_1–ξ_n the screws related to the revolute joints, M the contact point, L and L' the contact trajectories respectively on the fingertip and on the object, $\mathbf{e}_1$ and $\mathbf{e}_3$ respectively the tangent unit vector and the normal vector at the contact point M_i, as shown in Fig. 10.1.

The frame T on the fingertip w.r.t. the palm frame P can be obtained by the POE formula in terms of joint angles η_1–η_n as

$$\mathbf{g}_{PT} = e^{\xi_1' \eta_1} e^{\xi_2' \eta_2} \cdots e^{\xi_n' \eta_n} \mathbf{g}_{PT}(0) \tag{10.1}$$

where $\xi_i' = \mathrm{Ad}_{e^{\xi_1 \eta_1} e^{\xi_2 \eta_2} \cdots e^{\xi_{i-1} \eta_{i-1}}} \xi_i$, and ξ_i is the screw of the ith joint in the initial configuration. The matrix $\mathbf{g}_{ST}(0)$ is the initial location of the frame T.

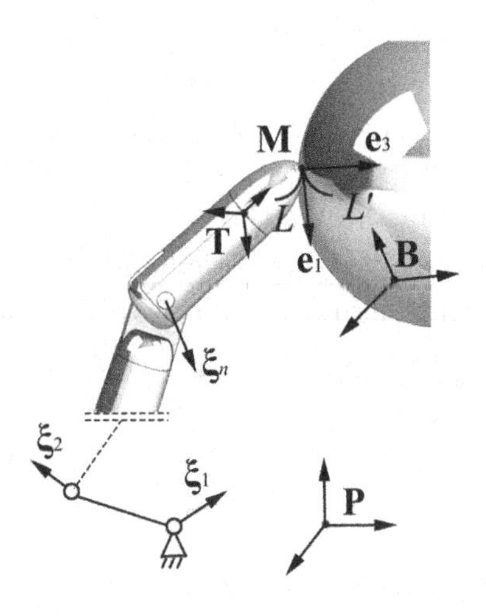

Fig. 10.1: In-hand manipulation with rolling contact

The twist of the frame T w.r.t. the frame P can be obtained as

$$\xi_{PT} = \dot{\mathbf{g}}_{ST}\mathbf{g}_{ST}^{-1} = \mathbf{J}(\eta)\dot{\eta} \tag{10.2}$$

where

$$\mathbf{J}(\eta) = [\xi_1', \ \xi_2', \ \dots \ \xi_n']$$
$$\dot{\eta} = [\dot{\eta}_1, \ \dot{\eta}_2, \ \dots, \ \dot{\eta}_n]^T$$

The homogeneous transformation matrix of the frame ($\mathbf{M}$, $\mathbf{e}_1$, $\mathbf{e}_2$, $\mathbf{e}_3$) w.r.t. the frame T is

$$\mathbf{g}_{TM} = \begin{bmatrix} \mathbf{E} & \mathbf{M} \\ \mathbf{0} & 1 \end{bmatrix} \tag{10.3}$$

where $\mathbf{E} = [\mathbf{e}_1^T, \ \mathbf{e}_2^T, \ \mathbf{e}_3^T]$ and $\mathbf{M}$ represents the position vector of the point M w.r.t. the frame T.

The differentiation of the matrix $\mathbf{E}$ w.r.t time t yields

$$\frac{d\mathbf{E}}{dt} = \frac{d\mathbf{E}}{ds}\frac{ds}{dt} = \sigma\mathbf{\Omega}\mathbf{E} \tag{10.4}$$

where the scalar $\sigma = ds/dt$ represents the rolling rate of the contact trajectory L on the fingertip, and the matrix $\mathbf{\Omega}$ contains the normal curvature k_n, geodesic curvature k_g, and geodesic torsion τ_g of L in the direction of $\mathbf{e}_1$:

$$\mathbf{\Omega} = \begin{bmatrix} 0 & k_n & k_g \\ -k_n & 0 & \tau_g \\ -k_g & -\tau_g & 0 \end{bmatrix}$$

The differentiation of the vector $\mathbf{M}$ w.r.t. time t gives

$$\frac{d\mathbf{M}}{dt} = \frac{d\mathbf{M}}{ds}\frac{ds}{dt} = \sigma\mathbf{e}_1 \tag{10.5}$$

Hence, the differentiation of $\mathbf{g}_{TM}$ w.r.t. time t can be obtained as

$$\dot{\mathbf{g}}_{TM} = \sigma\begin{bmatrix} \mathbf{\Omega}\mathbf{E} & \mathbf{e}_1^T \\ \mathbf{0} & 0 \end{bmatrix} \tag{10.6}$$

The twist of the frame ($\mathbf{M}$, $\mathbf{e}_1$, $\mathbf{e}_2$, $\mathbf{e}_3$) w.r.t. the frame T can be obtained as

$$\begin{aligned}\mathbf{V}_{TM} &= \dot{\mathbf{g}}_{TM}\mathbf{g}_{TM}^{-1} \\ &= \sigma\begin{bmatrix}\mathbf{\Omega E} & \mathbf{e}_1^T \\ \mathbf{0} & 0\end{bmatrix}\begin{bmatrix}\mathbf{E}^T & -\mathbf{E}^T\mathbf{M} \\ \mathbf{0} & 1\end{bmatrix} \\ &= \sigma\begin{bmatrix}\mathbf{\Omega} & \mathbf{e}_1^T - \mathbf{\Omega}\mathbf{M}^T \\ \mathbf{0} & 0\end{bmatrix}\end{aligned} \tag{10.7}$$

It follows that the twist of the frame ($\mathbf{M}$, $\mathbf{e}_1$, $\mathbf{e}_2$, $\mathbf{e}_3$) w.r.t. the frame T is

$$\xi_{TM} = \begin{bmatrix}\sigma\mathbf{e}_1 + \mathbf{M}\times\boldsymbol{\omega}_M \\ \boldsymbol{\omega}_M\end{bmatrix} \tag{10.8}$$

where

$$\boldsymbol{\omega}_M = \sigma(-\tau_g\mathbf{e}_1 + k_n\mathbf{e}_2 - k_g\mathbf{e}_3)$$

Note that the first three elements of ξ_{TM} in Eq. (10.8) represent the linear velocity of the point M w.r.t. the frame T. This velocity is the combined effects of velocity $\sigma\mathbf{e}_1^T$ induced by rolling and a velocity induced by rotation of the frame ($\mathbf{M}$, $\mathbf{e}_1$, $\mathbf{e}_2$, $\mathbf{e}_3$).

The angular velocity of the object frame B w.r.t. the frame ($\mathbf{M}$, $\mathbf{e}_1$, $\mathbf{e}_2$, $\mathbf{e}_3$) was obtained in Eq. (3.16) as

$$\boldsymbol{\omega}_B = \omega_1\mathbf{e}_1 + \omega_2\mathbf{e}_2 + \omega_3\mathbf{e}_3 \tag{10.9}$$

where $\omega_1 = -\sigma(\tau_g' - \tau_g)$, $\omega_2 = \sigma(k_n' - k_n)$, $\omega_3 = -\sigma(k_g' - k_g)$, and τ_g', k_n', k_g' are the geodesic torsion, normal curvature, and geodesic curvature of the contact curve L' in the direction of $\mathbf{e}_1$, respectively.

The twist of the object frame B w.r.t. the frame ($\mathbf{M}$, $\mathbf{e}_1$, $\mathbf{e}_2$, $\mathbf{e}_3$) is thus

$$\xi_{MB} = \begin{bmatrix}\mathbf{r}\times\boldsymbol{\omega}_B \\ \boldsymbol{\omega}_B\end{bmatrix} \tag{10.10}$$

where $\mathbf{r}$ represents the vector from the point M to B w.r.t. the frame ($\mathbf{M}$, $\mathbf{e}_1$, $\mathbf{e}_2$, $\mathbf{e}_3$).

It follows from Eqs. (10.2), (10.8), and (10.10) that the twist of the object frame B w.r.t. the palm frame P can be obtained as

$$\xi_{PB} = \xi_{PT} + \mathrm{Ad}_{\mathbf{g}_{PT}} \xi_{TM} + \mathrm{Ad}_{\mathbf{g}_{PM}} \xi_{MB} \tag{10.11}$$

This gives the forward instantaneous velocity of the object under the effects of joint rates and rolling contact.

10.3 Inverse Velocity Kinematics of In-Hand Manipulation with Rolling Contact

The inverse kinematics is to obtain the joint rates and the contact trajectories given the twist of the object. Suppose the surfaces of the fingertip is parameterized as $\mathbf{x}(u, v)$ and the surface of the object as $\mathbf{x}'(\alpha, \beta)$. Let $\mathbf{e}_u$ represent the tangent vector to the u-curve and $\mathbf{e}_\alpha$ the unit tangent vector to the α-curve. The common normal vector is $\mathbf{e}_3$ and the angle between $\mathbf{e}_u$ and $\mathbf{e}_\alpha$ is θ, as shown in Fig. 10.2.

A general rigid object has six DOFs, thus the number of finger joints is ideally three for a manipulation task in 3D space since rolling contact provides extra three terms to the system. In Chapter 6, these three terms have been defined as the rolling rate σ, the angle ϕ

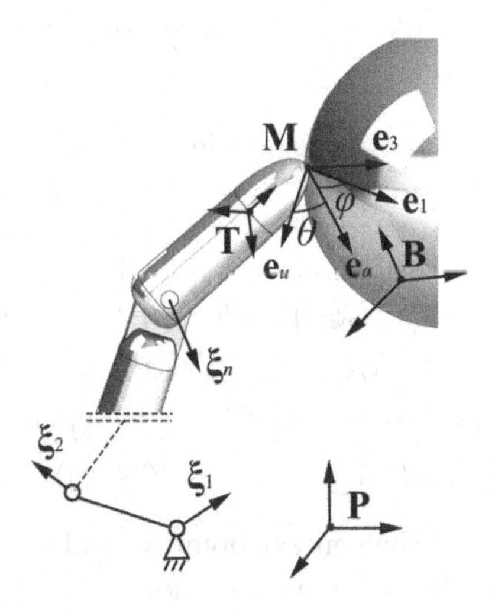

Fig. 10.2: In-hand manipulation with rolling contact

between the rolling direction $\mathbf{e}_1$ and $\mathbf{e}_\alpha$, and the complementary spin-speed ω_{spin}.

Hence, the problem of inverse kinematics amounts to obtaining the joint rates $\dot{\eta}_1$–$\dot{\eta}_n$ and the aforementioned three terms given the twist of the object ξ_{PB}.

Suppose the rolling direction $\mathbf{e}_1$ makes angle φ with $\mathbf{e}_\alpha$, as shown in Fig. 10.2. The vectors $\mathbf{e}_1$ and $\mathbf{e}_2$ can be expressed in the frame $(\mathbf{M}, \mathbf{e}_u, \mathbf{e}_v, \mathbf{e}_3)$ as

$$\begin{aligned} \mathbf{e}_1 &= \cos(\theta+\varphi)\mathbf{e}_u + \sin(\theta+\varphi)\mathbf{e}_v \\ \mathbf{e}_2 &= -\sin(\theta+\varphi)\mathbf{e}_u + \cos(\theta+\varphi)\mathbf{e}_v \end{aligned} \tag{10.12}$$

It follows from Eq. (2.45) that the normal curvature k_n, geodesic curvature k_g, and geodesic torsion τ_g in the direction of $\mathbf{e}_1$ can be obtained in terms of the curvatures of the u-curve and v-curve of the fingertip surface as

$$\begin{aligned} k_n &= k_{nu}\cos^2(\theta+\varphi) + \tau_{gu}\sin 2(\theta+\varphi) + k_{nv}\sin^2(\theta+\varphi) \\ k_g &= k_{gu}\cos(\theta+\varphi) + k_{gv}\sin(\theta+\varphi) + \frac{\omega_{\text{spin}}}{\sigma} \\ \tau_g &= \tau_{gu}\cos 2(\theta+\varphi) + \frac{1}{2}(k_{nv}-k_{nu})\sin 2(\theta+\varphi) \end{aligned} \tag{10.13}$$

The angular velocity $\boldsymbol{\omega}_M$ can be obtained from Eqs. (10.12) and (10.13) as

$$\boldsymbol{\omega}_M = F_1\mathbf{e}_u + F_2\mathbf{e}_v + F_3\mathbf{e}_3 \tag{10.14}$$

where

$$\begin{aligned} F_1 &= -\frac{\sigma}{2}\begin{pmatrix} (k_{nu}-k_{nv})\sin 3(\theta+\varphi) \\ +(k_{nu}+k_{nv})\sin(\theta+\varphi) + 2\tau_{gu}\cos(\theta+\varphi) \end{pmatrix} \\ F_2 &= \frac{\sigma}{2}\begin{pmatrix} (k_{nu}-k_{nv})\cos 3(\theta+\varphi) \\ +(k_{nu}+k_{nv})\cos(\theta+\varphi) + 2\tau_{gu}\sin(\theta+\varphi) \end{pmatrix} \\ F_3 &= -\sigma(k_{gu}\cos(\theta+\varphi) + k_{gv}\sin(\theta+\varphi)) - \omega_{\text{spin}} \end{aligned}$$

The coordinates of the contact point M and the curvatures of the u- and v-curves at the point M are known. Hence, the twist of the frame $(\mathbf{M}, \mathbf{e}_u, \mathbf{e}_v, \mathbf{e}_3)$ w.r.t. the frame T can be obtained as shown in Eq. (10.8).

The angular velocity of the object frame O w.r.t. the frame ($\mathbf{M}$, $\mathbf{e}_u$, $\mathbf{e}_v$, $\mathbf{e}_3$) was obtained in Eq. (6.7), and the coordinates of the origin of the frame B are known, giving the twist of the frame B as shown in Eq. (10.10).

Hence, it follows from Eq. (10.11) that the inverse kinematics is now formulated as a system of six nonlinear algebraic equations as

$$\mathbf{J}(\boldsymbol{\eta})\dot{\boldsymbol{\eta}} + \left[\, f_1(\sigma, \varphi, \omega_{\text{spin}}), \cdots, f_6(\sigma, \varphi, \omega_{\text{spin}})\right]^T = \boldsymbol{\xi}_{PB} \qquad (10.15)$$

where $f_i(\sigma, \varphi, \omega_{\text{spin}})$ are nonlinear scalar functions governed by the rolling motion.

In these six scalar equations, the number of unknowns is $n + 3$, namely $\dot{\eta}_1$ to $\dot{\eta}_n$, σ, φ, and ω_{spin}. Hence, it takes at least three joints on the finger to have full manipulability, with the other three inputs from rolling contact.

Suppose there are n joints on the finger with $3 \leq n \leq 6$. Taking the reciprocity product on the both sides of Eq. (10.15) by the reciprocal matrix $\mathbf{J}^R$ of the Jacobian $\mathbf{J}$ removes the joint velocity $\dot{\eta}_1$ to $\dot{\eta}_n$, yielding $6 - n$ scalar equations that only contains the three variables — σ, φ, and ω_{spin} — from the rolling contact, as shown in Eq. (10.16).

$$\mathbf{J}^R\Delta\left[\, f_1(\sigma, \varphi, \omega_{\text{spin}}), \ldots, f_6(\sigma, \varphi, \omega_{\text{spin}})\right]^T = \mathbf{J}^R\Delta\boldsymbol{\xi}_{PB} \qquad (10.16)$$

If $n = 3$, three scalar nonlinear equations will be generated, which have a finite number of solutions. If $n \geq 4$, an infinite number of solutions exist.

10.4 Examples of Application

10.4.1 *In-Hand Manipulation of a Planar Two-Fingered Robotic Hand*

In the first example, we present a planar two-fingered robotic hand to manipulate a disc of unit radius in 2D space, where each finger has two joints. Set up a frame ($\mathbf{P}$, $\mathbf{i}$, $\mathbf{j}$, $\mathbf{k}$) on the palm, a tool frame ($\mathbf{T}$, $\mathbf{i}'$, $\mathbf{j}'$, $\mathbf{k}'$), a Darboux frame ($\mathbf{M}$, $\mathbf{e}_1$, $\mathbf{e}_2$, $\mathbf{e}_3$), and an object frame B, as shown in Fig. 10.3.

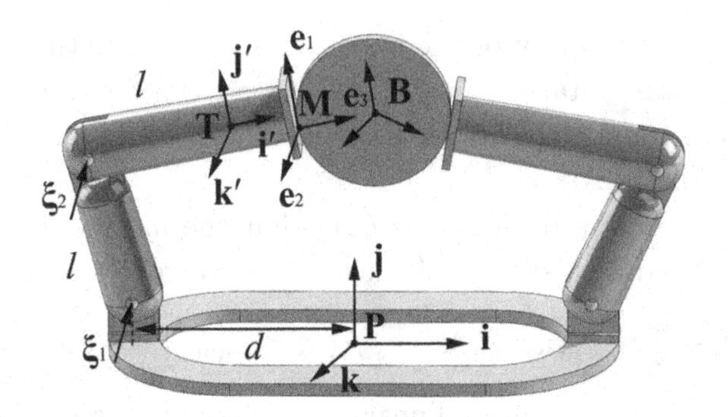

Fig. 10.3: A planar two-fingered robot hand manipulating a disc

Suppose that the origin of the palm frame ($\mathbf{P}$, $\mathbf{i}$, $\mathbf{j}$, $\mathbf{k}$) to the first joint of each finger is d, and the length of each finger's section is l and let η_1 and η_2 represent the two joint angles, respectively. The contact locus L on the end-effector of the finger is a straight line along the direction of $\mathbf{e}_1$, and the contact locus L' on the object (disc) is a circle of unit radius.

10.4.1.1 *Forward kinematics*

The tool frame ($\mathbf{T}$, $\mathbf{i}'$, $\mathbf{j}'$, $\mathbf{k}'$) w.r.t. the palm frame ($\mathbf{P}$, $\mathbf{i}$, $\mathbf{j}$, $\mathbf{k}$) can be obtained by the POE formula in joint space as

$$\begin{aligned}\mathbf{g}_{PT} &= e^{\xi_1\eta_1}e^{\xi_2\eta_2}\mathbf{g}_{PT}(0)\\ &= \begin{bmatrix}\cos\eta_{12} & -\sin\eta_{12} & 0 & l(\cos\eta_1+\cos\eta_{12})-d\\ \sin\eta_{12} & \cos\eta_{12} & 0 & l(\sin\eta_1+\sin\eta_{12})\\ 0 & 0 & 1 & 0\\ 0 & 0 & 0 & 1\end{bmatrix}\end{aligned}$$

where $\eta_{12}=\eta_1+\eta_2$. It follows from Eq. (10.2) that the twist ξ_{PT} is

$$\xi_{PT} = \begin{bmatrix} l\dot{\eta}_2\sin\eta_1\\ d\dot{\eta}_1+\dot{\eta}_2(d-l\sin\eta_1)\\ 0\\ 0\\ 0\\ \dot{\eta}_{12}\end{bmatrix}$$

The homogeneous transformation matrix of the frame ($\mathbf{M}$, $\mathbf{e}_1$, $\mathbf{e}_2$, $\mathbf{e}_3$) w.r.t. the frame ($\mathbf{T}$, $\mathbf{i}'$, $\mathbf{j}'$, $\mathbf{k}'$) is

$$\mathbf{g}_{TM} = \begin{bmatrix} 0 & 0 & 1 & M_x \\ 1 & 0 & 0 & M_y \\ 0 & 1 & 0 & 0 \\ 0 & 0 & 0 & 1 \end{bmatrix}$$

where M_x and M_y are the coordinates of the contact point M. The contact trajectory L on the fingertip is a straight line, thus the geodesic curvature k_g, normal curvature k_n, and geodesic torsion τ_g are 0. The twist ξ_{TM} can be obtained from Eq. (10.8) as

$$\xi_{TM} = [-\sigma \cos \eta_{12},\ \sigma \sin \eta_{12},\ 0,\ 0,\ 0,\ 0]^T$$

On the contact trajectory L', the geodesic curvature k'_g and geodesic torsion τ'_g are 0, and the normal curvature k'_n is 1. It follows from Eq. (10.9) that the angular velocity of the disc frame B w.r.t. the frame ($\mathbf{M}$, $\mathbf{e}_1$, $\mathbf{e}_2$, $\mathbf{e}_3$) is

$$\boldsymbol{\omega}_B = \sigma \mathbf{e}_2 = [0,\ 0,\ \sigma]^T$$

The twist of the object frame B w.r.t. the frame ($\mathbf{M}$, $\mathbf{e}_1$, $\mathbf{e}_2$, $\mathbf{e}_3$) can be obtained from Eq. (10.10) as

$$\xi_{MB} = [-\sigma,\ 0,\ 0,\ 0,\ \sigma,\ 0]^T$$

Note that the coordinates of the point B w.r.t. the frame ($\mathbf{M}$, $\mathbf{e}_1$, $\mathbf{e}_2$, $\mathbf{e}_3$) are $[0,\ 0,\ 1]$.

It follows from Eq. (10.11) that the twist of the object frame B w.r.t. the palm frame ($\mathbf{P}$, $\mathbf{i}$, $\mathbf{j}$, $\mathbf{k}$) can be obtained as

$$\xi_{PB} = \begin{bmatrix} A_1\sigma + l \sin \eta_1 \dot{\eta}_2 \\ A_2\sigma + d\dot{\eta}_1 + (d - l \cos \eta_1)\dot{\eta}_2 \\ 0 \\ 0 \\ \sigma + \dot{\eta}_{12} \end{bmatrix} \tag{10.17}$$

where

$$A_1 = M_y \cos\eta_{12} + M_x \sin\eta_{12} + l\sin\eta_{12} + l\sin\eta_1 + \sin\eta_{12}$$
$$A_2 = -M_y \sin\eta_{12} - M_x \cos\eta_{12} + l\cos\eta_{12} + l\sin\eta_1 - \cos\eta_{12} - d$$

This completes the forward kinematics.

10.4.1.2 *Inverse kinematics*

Suppose the angular velocity of the disc is $\omega\mathbf{k}$ and the coordinates of the disc center B is $(B_x,\ B_y)$ at time t. The twist of the disc w.r.t. the P-frame can be obtained as

$$\xi_{PB} = [B_y\omega,\ -B_x\omega,\ 0,\ 0,\ 0,\ \omega]^T \tag{10.18}$$

The problem of inverse kinematics amounts to equating the elements in Eq. (10.17) to the corresponding elements in Eq. (10.18), which generates a system of linear equations as

$$\begin{bmatrix} 0 & l\sin\eta_1 & A_1 \\ d & d - l\cos\eta_1 & A_2 \\ 1 & 1 & 1 \end{bmatrix} \begin{bmatrix} \dot{\eta}_1 \\ \dot{\eta}_2 \\ \sigma \end{bmatrix} = \begin{bmatrix} B_y\omega \\ -B_x\omega \\ \omega \end{bmatrix} \tag{10.19}$$

where A_1 and A_2 are the same as in Eq. (10.17). The variables $\dot{\eta}_1$, $\dot{\eta}_2$, and σ can be obtained straightforwardly.

10.4.2 *In-Hand Manipulation of the Three-Fingered Metahand*

We fixed the palm of the Metahand on a circle and equip it with three semi-ellipsoid fingertips. Set up a frame $(\mathbf{P},\ \mathbf{i},\ \mathbf{j},\ \mathbf{k})$ on the palm, a tool frame $(\mathbf{T},\ \mathbf{i}',\ \mathbf{j}',\ \mathbf{k}')$, a Darboux frame $(\mathbf{M},\ \mathbf{e}_1,\ \mathbf{e}_2,\ \mathbf{e}_3)$, and an object frame B. Suppose the origin of the frame P to the first joint of each finger is d, and the length of each finger section is l as shown in Fig. 10.4.

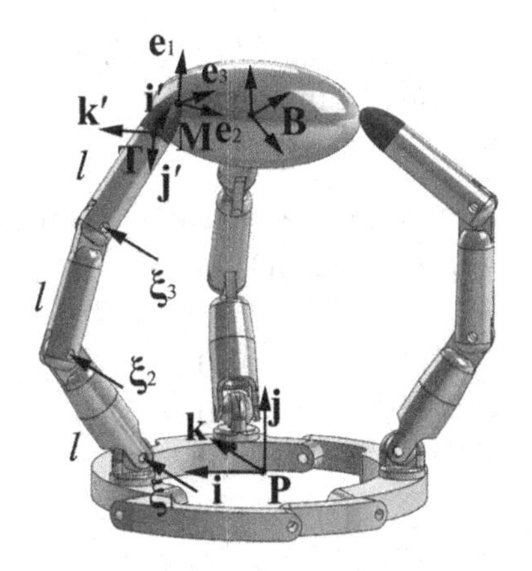

Fig. 10.4: Three-fingered in-hand manipulation of an ellipsoid

The fingertip, which is a semi-ellipsoid, can be parameterized as

$$\mathbf{x}(u, v) = [b_1 \cos v,\ a_1 \sin v \cos u,\ a_1 \sin v \sin u]$$
$$u \in [0, 2\pi), v \in \left[0, \frac{\pi}{2}\right)$$

The object is an ellipsoid which can be parameterized as

$$\mathbf{x}'(\alpha, \beta) = [b_2 \cos \beta,\ a_2 \sin \beta \cos \alpha,\ a_2 \sin \beta \sin \alpha]$$
$$\alpha \in [0, 2\pi), \beta \in [0, \pi)$$

10.4.2.1 *Forward kinematics*

The tool frame ($\mathbf{T}$, $\mathbf{i}'$, $\mathbf{j}'$, $\mathbf{k}'$) on the fingertip w.r.t. the palm frame ($\mathbf{P}$, $\mathbf{i}$, $\mathbf{j}$, $\mathbf{k}$) can be obtained by the POE formula as

$$\begin{aligned}\mathbf{g}_{PT} &= e^{\xi_1 \eta_1} e^{\xi_2 \eta_2} e^{\xi_3 \eta_3} \mathbf{g}_{PT}(0) \\ &= \begin{bmatrix} \cos \eta_{123} & -\sin \eta_{123} & 0 & l(\cos \eta_1 + \cos \eta_{12} + \cos \eta_{123}) + d \\ \sin \eta_{123} & \cos \eta_{123} & 0 & l(\sin \eta_1 + \sin \eta_{12} + \sin \eta_{123}) \\ 0 & 0 & 1 & 0 \\ 0 & 0 & 0 & 1 \end{bmatrix}\end{aligned}$$

where $\eta_{12} = \eta_1 + \eta_2$ and $\eta_{123} = \eta_1 + \eta_2 + \eta_3$. It follows from Eq. (10.2) that the twist ξ_{PT} can be obtained as

$$\xi_{PT} = \begin{bmatrix} l(\dot{\eta}_2 \sin \eta_1 + \dot{\eta}_3(\sin \eta_1 + \sin \eta_{12})) \\ -d\dot{\eta}_1 - \dot{\eta}_2(d + l \cos \eta_1) - \dot{\eta}_3(d + l \cos \eta_1 + l \cos \eta_{12}) \\ 0 \\ 0 \\ 0 \\ \dot{\eta}_{123} \end{bmatrix}$$

Suppose the contact trajectory L on the fingertip is a v-coordinate curve, which is circle, and the contact trajectory L' on the ellipsoidal object is an α-coordinate curve, which is an ellipse. The unit vectors $\mathbf{e}_1$ to $\mathbf{e}_3$ w.r.t. the frame $(\mathbf{T}, \mathbf{i}', \mathbf{j}', \mathbf{k}')$ are respectively:

$$\begin{aligned} \mathbf{e}_1 &= \frac{1}{D_1}[-b_1 \sin v,\ a_1 \cos u \cos v,\ a_1 \sin u \cos v] \\ \mathbf{e}_2 &= [0,\ \sin u,\ -\cos u] \\ \mathbf{e}_3 &= \frac{1}{D_1}[-a_1 \cos v,\ -b_1 \cos u \sin v,\ -b_1 \sin u \sin v] \end{aligned}$$

where $D_1 = \sqrt{a_1^2 \cos^2 v + b_1^2 \sin^2 v}$.

The normal curvature k_n, geodesic curvature k_g, and geodesic torsion τ_g of L in the direction of $\mathbf{e}_1$ can be obtained as

$$k_n = \frac{-a_1 b_1}{D_1^3}, \quad k_g = 0, \quad \tau_g = 0$$

The angular velocity of the frame $(\mathbf{M}, \mathbf{e}_1, \mathbf{e}_2, \mathbf{e}_3)$ w.r.t. the frame $(\mathbf{T}, \mathbf{i}', \mathbf{j}', \mathbf{k}')$ can be obtained from Eq. (10.8) as

$$\omega_M = \sigma(-\tau_g \mathbf{e}_1 + k_n \mathbf{e}_2 - k_g \mathbf{e}_3) = -\frac{\sigma a_1 b_1}{D_1^3} \mathbf{e}_2$$

It follows from Eq. (10.8) that the twist of the frame ($\mathbf{M}$, $\mathbf{e}_1$, $\mathbf{e}_2$, $\mathbf{e}_3$) w.r.t. the frame ($\mathbf{T}$, $\mathbf{i}'$, $\mathbf{j}'$, $\mathbf{k}'$) can be obtained as

$$\xi_{TM} = \begin{bmatrix} \sigma \mathbf{e}_1 + \mathbf{M} \times \omega_M \\ \omega_M \end{bmatrix} = \frac{\sigma}{D_1} \begin{bmatrix} \frac{a_1^2 b_1 \sin v}{D_1^2} \\ -a_1 b_1^2 \cos u \cos v \\ -a_1 b_1^2 \sin u \cos v \\ 0 \\ -a_1 b_1 \sin u \\ a_1 b_1 \cos u \end{bmatrix}$$

where $\mathbf{M} = [b_1 \cos v,\ a_1 \sin v \cos u,\ a_1 \sin v \sin u]$ w.r.t. the frame ($\mathbf{T}$, $\mathbf{i}'$, $\mathbf{j}'$, $\mathbf{k}'$).

The normal curvature k_n', geodesic curvature k_g', and geodesic torsion τ_g' of L' in the direction of $\mathbf{e}_1$ can be obtained as

$$k_n' = -\frac{a_2 b_2}{D_2^3}, \quad k_g' = 0, \quad \tau_g' = 0$$

where $D_2 = \sqrt{a_2^2 \cos^2 \beta + b_2^2 \sin^2 \beta}$.

It follows from Eq. (10.9) that the angular velocity of the object frame B w.r.t. the frame ($\mathbf{M}$, $\mathbf{e}_1$, $\mathbf{e}_2$, $\mathbf{e}_3$) can be obtained as

$$\omega_B = \sigma((\tau_g' - \tau_g)\mathbf{e}_1 + (k_n' - k_n)\mathbf{e}_2 + (k_g' - k_g)\mathbf{e}_3) = \sigma\left(-\frac{a_2 b_2}{D_2^3} + \frac{a_1 b_1}{D_1^3}\right)\mathbf{e}_2$$

It follows from Eq. (10.10) that the twist of the object frame B w.r.t. the frame ($\mathbf{M}$, $\mathbf{e}_1$, $\mathbf{e}_2$, $\mathbf{e}_3$) can be obtained as

$$\xi_B = \begin{bmatrix} \mathbf{B} \times \omega \\ \omega_B \end{bmatrix} = \begin{bmatrix} -a_2^2 b_2 \left(\frac{a_2 b_2}{D_2^3} + \frac{a_1 b_1}{D_1^3}\right) \\ 0 \\ 0 \\ 0 \\ -\frac{a_2 b_2}{D_2^3} - \frac{a_1 b_1}{D_1^3} \\ 0 \end{bmatrix}$$

where $\mathbf{B} = \frac{1}{D_2^2}(\sin\beta \cos\beta (a_2^2 - b_2^2)\mathbf{e}_2 - a_2 b_2 \mathbf{e}_3)$ w.r.t. the frame ($\mathbf{M}$, $\mathbf{e}_1$, $\mathbf{e}_2$, $\mathbf{e}_3$).

The twist of the object frame B w.r.t. the palm frame ($\mathbf{P}$, $\mathbf{i}$, $\mathbf{j}$, $\mathbf{k}$) can be obtained from Eq. (10.11) straightforwardly.

10.4.2.2 *Inverse kinematics*

We use a numerical example to demonstrate the inverse kinematics. Closed-form solution only exists in certain circumstances.

Suppose the desired twist of the object is $\xi_{PB} = [2,\ -0.5,\ -1.5,\ 1,\ 1,\ 1]$ and the structural parameters are: $d = 1.5$, $l = 1$, $a_1 = 0.2$, $b_1 = 0.15$, $a_2 = 0.8$, $b_2 = 0.6$. The robot hand is in the following configuration:

$$\eta_1 = \frac{\pi}{3}, \quad \eta_2 = \frac{\pi}{3}, \quad \eta_3 = \frac{\pi}{6}$$
$$u_0 = \frac{\pi}{4}, \quad v_0 = \frac{3\pi}{4}$$
$$\alpha_0 = \frac{\pi}{4}, \quad \beta = \frac{\pi}{4}$$

It follows from Eq. (10.2) that the homogeneous transformation matrix $\mathbf{g}_{PT}$ can be obtained as

$$\mathbf{g}_{PT} = \begin{bmatrix} -0.5 & -0.866 & 0 & 0.634 \\ 0.866 & -0.5 & 0 & 2.232 \\ 0 & 0 & 1 & 0 \\ 0 & 0 & 0 & 1 \end{bmatrix}$$

The Jacobian of the finger joints is

$$\begin{bmatrix} 0 & 0.866 & 1.732 \\ -1.5 & -2.0 & -1.5 \\ 0 & 0 & 0 \\ 0 & 0 & 0 \\ 0 & 0 & 0 \\ 1 & 1 & 1 \end{bmatrix}$$

The coordinates of the contact point M w.r.t. the frame ($\mathbf{T}$, $\mathbf{i}'$, $\mathbf{j}'$, $\mathbf{k}'$) is

$$\mathbf{M} = [0.106,\ 0.10,\ 0.10]^T$$

The unit tangent vectors $\mathbf{e}_u$, $\mathbf{e}_v$, and $\mathbf{e}_3$ at the point M of the fingertip ellipsoid can be obtained as

$$\begin{aligned}\mathbf{e}_u &= [0,\ -0.710,\ 0.710]\\ \mathbf{e}_v &= [-0.600,\ 0.570,\ 0.570]\\ \mathbf{e}_3 &= [-0.809,\ -0.426,\ -0.426]\end{aligned}$$

The curvatures of the u-curve and v-curve of the fingertip ellipsoid can be obtained as

$$\begin{aligned}&k_{nu} = -4.234, \quad k_{nv} = -5.431, \quad k_{gu} = -5.657,\\ &k_{gv} = 0, \quad \tau_{gu} = \tau_{gv} = 0\end{aligned}$$

The unit tangent vectors $\mathbf{e}_\alpha$ and $\mathbf{e}_\beta$ of the object at the contact point M can be obtained as

$$\begin{aligned}\mathbf{e}_\alpha &= [0,\ -0.866,\ 0.500]\\ \mathbf{e}_\beta &= [-0.756,\ 0.327,\ 0.567]\end{aligned}$$

The curvatures of the α and β curves of the object can be obtained as

$$\begin{aligned}&k_{n\alpha} = 1.455, \quad k_{n\beta} = 2.493, \quad k_{g\alpha} = -1.260,\\ &k_{g\beta} = 0, \quad \tau_{g\alpha} = \tau_{g\beta} = 0\end{aligned}$$

The angle θ between $\mathbf{e}_u$ and $\mathbf{e}_\alpha$ is 0.246 rad. Suppose that the unit vector $\mathbf{e}_1$ of the rolling direction makes an angle φ with $\mathbf{e}_\alpha$. Then the vectors $\mathbf{e}_1$ and $\mathbf{e}_2$ can be obtained from Eq. (10.12) as

$$\mathbf{e}_1 = \begin{bmatrix} -0.756\sin(\varphi + 0.262)\\ -0.866\cos(\varphi + 0.262) + 0.327\sin(\varphi + 0.262)\\ 0.500\cos(\varphi + 0.262) + 0.567\sin(\varphi + 0.262)\end{bmatrix}^T$$

$$\mathbf{e}_2 = \begin{bmatrix} -0.756\cos(\varphi + 0.262)\\ 0.866\sin(\varphi + 0.262) + 0.327\cos(\varphi + 0.262)\\ -0.500\sin(\varphi + 0.262) + 0.567\cos(\varphi + 0.262)\end{bmatrix}^T$$

It follows from Eq. (10.14) that the angular velocity of the frame $(\mathbf{M}, \mathbf{e}_1, \mathbf{e}_2, \mathbf{e}_3)$ in terms of the rolling rate σ, rolling direction φ, and compensatory spin rate ω_{spin} can be obtained as

$$\boldsymbol{\omega}_M = \begin{bmatrix} \sigma A_1 + 0.809\omega_{\text{spin}} \\ \sigma A_2 + 0.426\omega_{\text{spin}} \\ \sigma A_3 + 0.426\omega_{\text{spin}} \end{bmatrix}$$

where

$$\begin{aligned} A_1 &= -1.008\cos^3\varphi + 1.008\cos^2\varphi\sin\varphi + 0.168\sin\varphi - 0.812\cos\varphi \\ A_2 &= 2.138\cos^3\varphi + 0.238\cos^2\varphi\sin\varphi - 2.034\sin\varphi - 7.45\cos\varphi \\ A_3 &= -0.238\cos^3\varphi - 2.138\cos^2\varphi\sin\varphi - 3.898\sin\varphi + 5.167\cos\varphi \end{aligned}$$

It follows from Eq. (10.8) that the twist ξ_{TM} can be obtained straightforwardly with $\mathbf{M} = [0.106, 0.10, 0.10]^T$.

The angular velocity of the object w.r.t. the frame ($\mathbf{M}$, $\mathbf{e}_1$, $\mathbf{e}_2$, $\mathbf{e}_3$) can be obtained from Eq. (10.10) as

$$\boldsymbol{\omega}_O = \begin{bmatrix} -\sigma(1.820\cos\varphi + 7.642\sin\varphi) \\ \sigma(5.490\cos\varphi - 1.784\sin\varphi) \\ -\sigma(4.23\cos\varphi - 1.50\sin\varphi) + \omega_{\text{spin}} \end{bmatrix}$$

The coordinates of the center B w.r.t. the frame ($\mathbf{M}$, $\mathbf{e}_1$, $\mathbf{e}_2$, $\mathbf{e}_3$) are

$$\mathbf{B} = \begin{bmatrix} -0.282\sin(\varphi + 0.262) \\ 0.135\sin(\varphi + 0.262) \\ 0.4611 \end{bmatrix}$$

The twist of ξ_{MB} can be obtained from Eq. (10.10) straightforwardly.

It follows from Eq. (10.15) that the governing equation of the inverse kinematics in terms of the joint rates $\dot{\eta}_1$, $\dot{\eta}_2$, $\dot{\eta}_3$, rolling rate σ, rolling direction φ, and compensatory spin rate ω_{spin} can be obtained as

$$\begin{bmatrix} 0 & 0.866 & 1.732 \\ -1.5 & -2.0 & -1.5 \\ 0 & 0 & 0 \\ 0 & 0 & 0 \\ 0 & 0 & 0 \\ 1 & 1 & 1 \end{bmatrix} \begin{bmatrix} \dot{\eta}_1 \\ \dot{\eta}_2 \\ \dot{\eta}_3 \end{bmatrix} + \begin{bmatrix} f_1(\sigma, \varphi, \omega_{\text{spin}}) \\ f_2(\sigma, \varphi, \omega_{\text{spin}}) \\ f_3(\sigma, \varphi, \omega_{\text{spin}}) \\ f_4(\sigma, \varphi, \omega_{\text{spin}}) \\ f_5(\sigma, \varphi, \omega_{\text{spin}}) \\ f_6(\sigma, \varphi, \omega_{\text{spin}}) \end{bmatrix} = \xi_{PO} \quad (10.20)$$

Taking the reciprocal product on both sides of Eq. (10.20) with the reciprocal matrix $\mathbf{J}^R$ of the joint Jacobian $\mathbf{J}$ eliminates the joint rates $\dot{\eta}_1$, $\dot{\eta}_2$, and $\dot{\eta}_3$, yielding three nonlinear algebraic equations in terms of the rolling rate σ, rolling direction φ, and compensatory spin rate ω_{spin}. It is straightforward to obtain the numerical solution to this system of nonlinear equations and we skip the detailed calculations here.

10.5 Conclusion

We have extended the results in Chapter 8 to the kinematics of in-hand manipulation with rolling contact between the fingertips and the object. Rolling contact adds complexity into singularity analysis, which is evidenced in the example of in-hand manipulation by a planar two-fingered robotic hand. We have then demonstrated the integration of rolling kinematics into in-hand manipulation of the Metahand.

Appendix

We include the necessary elements related to in-hand manipulation in the Appendix for the completeness of this book. Details can be found in (Murray *et al.*, 1994; Selig, 2005).

A.1 Lie Group SE(3) and Lie Algebra se(3)

Consider a rigid body moving in free space. Assume a space reference frame $(\mathbf{O}, \mathbf{i}, \mathbf{j}, \mathbf{k})$, which is denoted by $\{S\}$, fixed in space and a body frame $(\mathbf{O}', \mathbf{i}', \mathbf{j}', \mathbf{k}')$, which is denoted by $\{T\}$, fixed to the body, as shown in Fig. A.1.

At any instance, the configuration (position and orientation) of the rigid body can be described by a 4×4 homogeneous transformation matrix $\mathbf{g}_{ST}$, which corresponds to the displacement from the

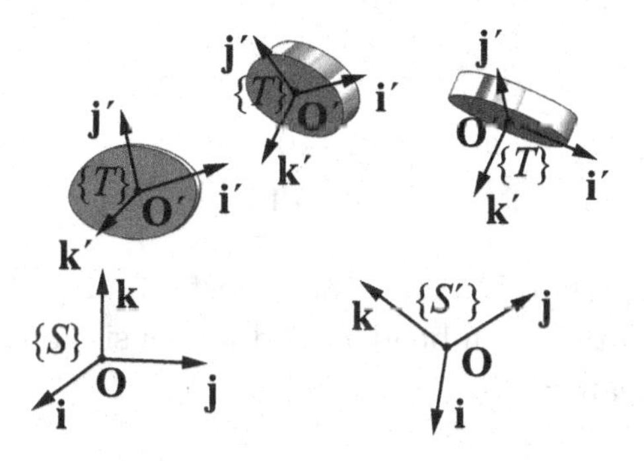

Fig. A.1: The fixed-space frame and the frame attached to the moving rigid object

reference frame $\{S\}$ to the body frame $\{T\}$. This matrix can be written as

$$\mathbf{g}_{ST} = \begin{bmatrix} \mathbf{R} & \mathbf{p} \\ \mathbf{0} & 1 \end{bmatrix} \tag{A.1}$$

where $\mathbf{p}$ is a vector representing the position vector from $\mathbf{O}$ to $\mathbf{O}'$ referenced w.r.t. the frame $\{S\}$ and $\mathbf{R}$ is a matrix representing the rotation of frame $\{T\}$ w.r.t. the frame $\{S\}$.

The matrix $\mathbf{R}$ has the properties that $\det \mathbf{R} = 1$ and $\mathbf{R}^T\mathbf{R} = \mathbf{I}$. The set of all 3×3 matrices that have these two properties is denoted SO(3), where SO abbreviates special orthogonal. The matrix $\mathbf{g}_{ST}$ is called the homogeneous representation of SE(3), where SE abbreviates special euclidean. Both SO(3) and SE(3) are Lie groups under the operation of matrix multiplication.

On any Lie group, the tangent space at the group identity has the structure of a Lie algebra. The Lie algebra of SO(3) and SE(3), denoted by so(3) and se(3), respectively, are given by

$$\begin{aligned} \text{so}(3) &= \left\{ \hat{\boldsymbol{\omega}} \middle| \hat{\boldsymbol{\omega}} \in \mathbb{R}^{3\times 3}, \hat{\boldsymbol{\omega}}^T = -\hat{\boldsymbol{\omega}} \right\} \\ \text{se}(3) &= \left\{ \hat{\xi} = \begin{bmatrix} \hat{\boldsymbol{\omega}} & \mathbf{v} \\ \mathbf{0} & 0 \end{bmatrix} \middle| \hat{\boldsymbol{\omega}} \in \text{so}(3), \mathbf{v} \in \mathbb{R}^3 \right\} \end{aligned} \tag{A.2}$$

where $\boldsymbol{\omega}$ and $\mathbf{v}$ are three-dimensional column vectors and the $\hat{}$ operator maps $\boldsymbol{\omega} = [\omega_1,\ \omega_2,\ \omega_3]^T$ to a 3×3 skew-symmetric matrix:

$$\hat{\boldsymbol{\omega}} - \begin{bmatrix} 0 & -\omega_3 & \omega_2 \\ \omega_3 & 0 & -\omega_1 \\ -\omega_2 & \omega_1 & 0 \end{bmatrix}$$

An element of se(3) is referred as a twist, or an infinitesimal generator of SE(3). A twist can be identified with a six-dimensional vector with the $\vee$ operator:

$$\begin{bmatrix} \hat{\boldsymbol{\omega}} & \mathbf{v} \\ \mathbf{0} & 0 \end{bmatrix}^{\vee} = \begin{bmatrix} \mathbf{v} \\ \boldsymbol{\omega} \end{bmatrix} = \xi \tag{A.3}$$

We call ξ the twist coordinates of $\hat{\xi}$. The operator $\wedge$ maps a twist to a matrix in se(3):

$$\xi^\wedge = \begin{bmatrix} \mathbf{v} \\ \omega \end{bmatrix}^\wedge = \begin{bmatrix} \hat{\omega} & \mathbf{v} \\ \mathbf{0} & 0 \end{bmatrix} \tag{A.4}$$

Given a curve

$$\mathbf{g}_{ST}(t) : [-a, a] \to \mathrm{SE}(3), \mathbf{g}_{ST}(t) = \begin{bmatrix} \mathbf{R}(t) & \mathbf{p}(t) \\ \mathbf{0} & 1 \end{bmatrix} \tag{A.5}$$

The space velocity $\hat{\mathbf{V}}^s_{ST} \in$ se(3) is defined to be

$$\hat{\mathbf{V}}^s_{ST} = \dot{\mathbf{g}}_{ST}\mathbf{g}_{ST}^{-1} = \begin{bmatrix} \dot{\mathbf{R}}\mathbf{R}^{-1} & -\mathbf{R}^{-1}\mathbf{p} + \dot{\mathbf{p}} \\ \mathbf{0} & 0 \end{bmatrix} \tag{A.6}$$

The space velocity in Eq. (A.5) can be identified with a twist ξ_{ST} by the $\vee$ operator.

Let $\{S'\}$ be another space frame, as shown in Fig. A.1, where frame $\{S\}$ can be obtained from $\{S'\}$ by a transformation $\mathbf{g}_{S'S}$. It follows that $\mathbf{g}_{S'T} = \mathbf{g}_{S'S}\mathbf{g}_{ST}$. The space velocity of frame $\{T\}$ w.r.t. frame $\{S'\}$ can be obtained as

$$\begin{aligned} \hat{\mathbf{V}}^s_{S'T} &= \dot{\mathbf{g}}_{S'T}\mathbf{g}_{S'T}^{-1} = \frac{d}{dt}\left(\mathbf{g}_{S'S}\mathbf{g}_{ST}\right)\left(\mathbf{g}_{S'S}\mathbf{g}_{ST}\right)^{-1} \\ &= \left(\dot{\mathbf{g}}_{S'S}\mathbf{g}_{ST} + \mathbf{g}_{S'S}\dot{\mathbf{g}}_{ST}\right)\mathbf{g}_{ST}^{-1}\mathbf{g}_{S'S}^{-1} \\ &= \hat{\mathbf{V}}^s_{S'S} + \mathbf{g}_{S'S}\hat{\mathbf{V}}^s_{ST}\mathbf{g}_{S'S}^{-1} \\ &= \hat{\mathbf{V}}^s_{S'S} + \mathrm{Ad}_{\mathbf{g}_{S'S}}\hat{\mathbf{V}}^s_{ST} \end{aligned} \tag{A.7}$$

If each space velocity is identified with a twist, the adjoint transformation associated with $\mathbf{g}_{S'S} = \left[\begin{smallmatrix} \mathbf{R}' & \mathbf{p}' \\ 0 & 1 \end{smallmatrix}\right]$ is given as

$$\mathrm{Ad}_{\mathbf{g}_{S'S}} = \begin{bmatrix} \mathbf{R}' & \hat{\mathbf{p}}'\mathbf{R}' \\ \mathbf{0} & \mathbf{R}' \end{bmatrix} \tag{A.8}$$

Given $\hat{\xi} \in se(3)$ and $\theta \in \mathbb{R}$, the exponential of $\hat{\xi}\theta$ is an element of SE(3). When $\omega = \mathbf{0}$, the map is given as

$$e^{\hat{\xi}\theta} = \begin{bmatrix} \mathbf{I} & \theta\mathbf{v} \\ \mathbf{0} & 1 \end{bmatrix} \tag{A.9}$$

When $\omega \neq \mathbf{0}$, we can normalize $\hat{\xi}$ in such a way that $\|\omega\| = 1$. Then the map is given as

$$e^{\hat{\xi}\theta} = \begin{bmatrix} e^{\hat{\omega}\theta} & (\mathbf{I} - e^{\hat{\omega}\theta})(\omega \times \mathbf{v}) + \theta\omega\omega^T\mathbf{v} \\ \mathbf{0} & 1 \end{bmatrix} \tag{A.10}$$

where

$$e^{\hat{\omega}\theta} = \mathbf{I} + \hat{\omega}\sin\theta + \hat{\omega}^2(1 - \cos\theta)$$

The exponential map for a twist gives the relative motion of a rigid body. Let $\mathbf{g}_{ST}(0)$ represent the initial configuration of a rigid body relative to a frame S, then the final configuration, still referenced w.r.t. the frame S, is given by

$$\mathbf{g}_{ST}(\theta) = e^{\hat{\xi}\theta}\mathbf{g}_{ST}(0) \tag{A.11}$$

A.2 Screw System and its Reciprocal Screw System

A screw is a geometric description of a twist. The geometric meaning of a screw is expressed in the famous Chasles' theorem: Every rigid body motion can be realized by a rotation about an axis combined with a translation parallel to that axis.

A screw ξ consists of an axis l, a pitch h, and a magnitude M. The axis l is a directed line in the direction of ω through the point $\frac{\omega^T \times \mathbf{v}}{\omega^T \cdot \omega}$ when $\omega \neq \mathbf{0}$ or through the origin when $\omega \neq \mathbf{0}$. The pitch h is the ratio of translational motion to rotational motion and $h = \frac{\omega^T \cdot \mathbf{v}}{\omega^T \cdot \omega}$. The magnitude M is the net rotation $\|\omega\|$ when $\omega \neq \mathbf{0}$ or net translation $\|\mathbf{v}\|$ when $\omega = \mathbf{0}$.

We can associate rigid motions due to a revolute joint to a unit twist where $\|\omega\| = 1$ and those due to a prismatic joint to a unit twist where $\omega = 0$ and $\|\mathbf{v}\| = 1$. It follows that $\mathbf{g}(\theta) = e^{\xi\theta}$, where θ is the amount of rotation or translation.

The reciprocal product of two screws $\xi_1 = [\mathbf{v}_1^T, \omega_1^T]^T$ and $\xi_2 = [\mathbf{v}_2^T, \omega_2^T]^T$ using twist coordinate is defined as

$$\xi_1 \Delta \xi_2 = \mathbf{v}_1^T \omega_2 + \mathbf{v}_2^T \omega_1 \tag{A.12}$$

Two reciprocal screws ξ_1 and ξ_2 are reciprocal if the reciprocal product of these two screws is 0.

We call a set of screws $\xi_1, \xi_2, \ldots, \xi_n$ a system of screws, which forms a linear space over the reals. We call the set of screws that are reciprocal to a given system of screws as the reciprocal screw system, which defines a linear subspace of twists.

A.3 Product-of-Exponentials Formula for Manipulator Kinematics

The forward kinematics of an open-chain n-DOF manipulator can be described by the product-of-exponentials (POE) formula. Let S be a frame attached to the base of the manipulator and T be a frame attached to the last link of the manipulator. Define the reference configuration of the manipulator to be the configuration of the manipulator corresponding to $\theta = 0$.

Let $\mathbf{g}_{ST}(0)$ represent the rigid body transformation between T and S when the manipulator is in its reference configuration. For each joint, construct a twist ξ_i that corresponds to the screw motion for the ith joint with all other joint angles fixed at $\theta_j = 0$. For a revolute joint, the twist ξ_i has the following form:

$$\xi_i = \begin{bmatrix} -\omega_i \times q_i \\ \omega_i \end{bmatrix} \tag{A.13}$$

where $\omega_i \in \mathbb{R}^3$ a unit vector in the direction of the twist axis and $q_i \in \mathbb{R}^3$ is any point on the axis.

Specifically, for a prismatic joint,

$$\xi_i = \begin{bmatrix} \mathbf{v}_i \\ 0 \end{bmatrix} \tag{A.14}$$

where $\omega_i \in \mathbb{R}^3$ is a unit vector pointing in the direction of translation.

Combining the individual joint motions, the forward kinematics map can be given as

$$\mathbf{g}_{ST}(\theta) = e^{\hat{\xi}_1\theta_1} e^{\hat{\xi}_2\theta_2} \ldots e^{\hat{\xi}_n\theta_n} \mathbf{g}_{ST}(0) \tag{A.15}$$

where $e^{\hat{\xi}_i\theta_i}$ is given in Eq. (A.10).

Let ξ_{ST} represent the twist coordinates of the end-effector w.r.t. the spatial frame S. It follows that

$$\begin{aligned}
\hat{\xi}_{ST} &= \sum_i \frac{\partial \mathbf{g}_{ST}(\theta_i)}{\partial \theta_i} \mathbf{g}_{ST}^{-1}(\theta) \\
&= ((\xi_1\dot{\theta}_1) e^{\hat{\xi}_1\theta_1} e^{\hat{\xi}_2\theta_2} \ldots e^{\hat{\xi}_n\theta_n} \mathbf{g}_{ST}(0))(\mathbf{g}_{ST}^{-1}(0) e^{-\hat{\xi}_n\theta_n} \cdot e^{-\hat{\xi}_2\theta_2} e^{-\hat{\xi}_1\theta_1}) \\
&\quad + (e^{\hat{\xi}_1\theta_1}(\xi_2\dot{\theta}_2) e^{\hat{\xi}_2\theta_2} \ldots e^{\hat{\xi}_n\theta_n} \mathbf{g}_{ST}(0)) \\
&\quad (\mathbf{g}_{ST}^{-1}(0) e^{-\hat{\xi}_n\theta_n} \cdot e^{-\hat{\xi}_2\theta_2} e^{-\hat{\xi}_1\theta_1}) \\
&\quad \vdots \\
&\quad + (e^{\hat{\xi}_1\theta_1} e^{\hat{\xi}_2\theta_2} \ldots (\xi_n\dot{\theta}_n) e^{\hat{\xi}_n\theta_n} \mathbf{g}_{ST}(0)) \\
&\quad (\mathbf{g}_{ST}^{-1}(0) e^{-\hat{\xi}_n\theta_n} \cdot e^{-\hat{\xi}_2\theta_2} e^{-\hat{\xi}_1\theta_1}) \\
&= \xi_1\dot{\theta}_1 + e^{\hat{\xi}_1\theta_1} \xi_2\dot{\theta}_2 e^{-\hat{\xi}_1\theta_1} + \cdots \\
&\quad + e^{\hat{\xi}_1\theta_1} e^{\hat{\xi}_2\theta_2} \ldots e^{\hat{\xi}_{n-1}\theta_{n-1}} \xi_n\dot{\theta}_n e^{-\hat{\xi}_{n-1}\theta_{n-1}} \ldots e^{-\hat{\xi}_2\theta_2} e^{-\hat{\xi}_1\theta_1}
\end{aligned} \tag{A.16}$$

Equation (A.16) can be written in matrix form as

$$\xi_{ST} = \mathbf{J}_{ST}(\theta)\dot{\theta} \tag{A.17}$$

where

$$\begin{aligned}
\mathbf{J}_{ST}(\theta) &= [\xi_1, \xi_2', \ldots, \xi_n'] \\
\xi_i' &= \mathrm{Ad}_{e^{\hat{\xi}_1\theta_1} \ldots e^{\hat{\xi}_{i-1}\theta_{i-1}}} \xi_i
\end{aligned}$$

This completes the use of POE to formulate the forward kinematics of a manipulator.

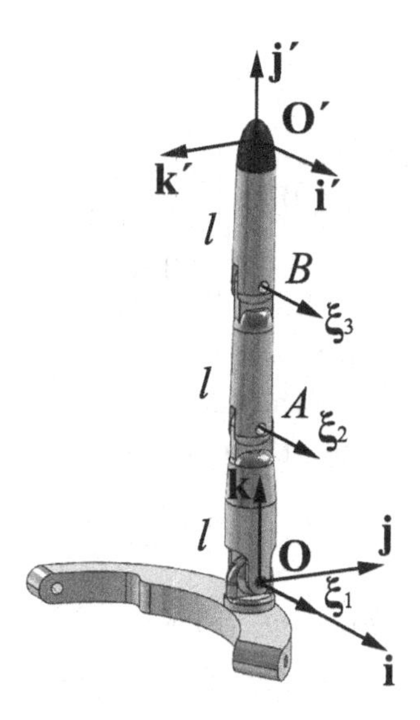

Fig. A.2: A $3R$ planar manipulator and its joint screws

The inverse kinematics is to obtain the joint angles given a desired configuration $\mathbf{g}_{ST}$ or a twist ξ_{ST}.

Example A.1. Obtain the forward and inverse kinematics of the planar $3R$ manipulator, as shown in Fig. A.2.

Solution:

Forward Position and Velocity Kinematics

We set up a space frame S represented by $(\mathbf{O}, \mathbf{i}, \mathbf{j}, \mathbf{k})$ such that the $\mathbf{k}$-axis is perpendicular to the circular palm link and the $\mathbf{j}$-axis is pointing outward of the circular palm link, as shown in Fig. A.2.

We choose the reference configuration of the $3R$ manipulator such that every link of the $3R$ manipulator is perpendicular to the circular palm link. We set up a tool frame T represented by $(\mathbf{O}', \mathbf{i}', \mathbf{j}', \mathbf{k}')$ such that the $\mathbf{j}'$-axis is parallel to the $\mathbf{k}$-axis and the $\mathbf{i}'$-axis is parallel to the $\mathbf{i}$-axis, as shown in Fig. A.2.

Forward kinematics is to obtain the position and orientation of the tool frame T. Suppose the three links have the same length l. Then the homogeneous matrix of the reference configuration of the tool frame T is

$$\mathbf{g}_{ST}(0) = \begin{bmatrix} 1 & 0 & 0 & 0 \\ 0 & 0 & -1 & 0 \\ 0 & 1 & 0 & 3l \\ 0 & 0 & 0 & 1 \end{bmatrix} \tag{A.18}$$

We associate a screw ξ_i to the ith joint of the manipulator and let θ_i denote the joint angle of the screw ξ_i referenced w.r.t. the frame $(\mathbf{O}, \mathbf{i}, \mathbf{j}, \mathbf{k})$. The axes of the three screws are in the direction of the $\mathbf{i}$-axis. It follows that

$$\omega_1 = \omega_2 = \omega_3 = \begin{bmatrix} 1 \\ 0 \\ 0 \end{bmatrix}$$

It is clear that the point O is on the screw ξ_1, the point A on ξ_2, and B on ξ_3. The position vectors of the points O, A, and B referenced w.r.t. the frame $(\mathbf{O}, \mathbf{i}, \mathbf{j}, \mathbf{k})$ are respectively,

$$\mathbf{O} = \begin{bmatrix} 0 \\ 0 \\ 0 \end{bmatrix}, \quad \mathbf{A} = \begin{bmatrix} 0 \\ 0 \\ l \end{bmatrix}, \quad \mathbf{B} = \begin{bmatrix} 0 \\ 0 \\ 2l \end{bmatrix}$$

The linear velocities of the points O, A, and B can be obtained as

$$v_1 = -\omega_1 \times \mathbf{O} = \begin{bmatrix} 0 \\ 0 \\ 0 \end{bmatrix}$$

$$v_2 = -\omega_2 \times \mathbf{A} = \begin{bmatrix} 0 \\ l \\ 0 \end{bmatrix}$$

$$v_3 = -\omega_3 \times \mathbf{B} = \begin{bmatrix} 0 \\ 2l \\ 0 \end{bmatrix}$$

The screws ξ_1–ξ_3 can be obtained as

$$\xi_1 = \begin{bmatrix} -\omega_1 \times \mathbf{O} \\ \omega_1 \end{bmatrix} = \begin{bmatrix} \mathbf{v}_1 \\ \omega_1 \end{bmatrix} = \begin{bmatrix} 0 \\ 0 \\ 0 \\ 1 \\ 0 \\ 0 \end{bmatrix}$$

$$\xi_2 = \begin{bmatrix} -\omega_2 \times \mathbf{A} \\ \omega_2 \end{bmatrix} = \begin{bmatrix} \mathbf{v}_2 \\ \omega_2 \end{bmatrix} = \begin{bmatrix} 0 \\ l \\ 0 \\ 1 \\ 0 \\ 0 \end{bmatrix} \tag{A.19}$$

$$\xi_3 = \begin{bmatrix} -\omega_3 \times \mathbf{B} \\ \omega_2 \end{bmatrix} = \begin{bmatrix} \mathbf{v}_3 \\ \omega_3 \end{bmatrix} = \begin{bmatrix} 0 \\ 2l \\ 0 \\ 1 \\ 0 \\ 0 \end{bmatrix}$$

The homogeneous matrices of the exponential of these screws are given by

$$e^{\hat{\xi}_1\theta_1} = \begin{bmatrix} e^{\hat{\omega}_1\theta_1} & (\mathbf{I} - e^{\hat{\omega}_1\theta_1})(\omega_1 \times \mathbf{v}_1) \\ \mathbf{0} & 1 \end{bmatrix} = \begin{bmatrix} 1 & 0 & 0 & 0 \\ 0 & \cos\theta_1 & -\sin\theta_1 & 0 \\ 0 & \sin\theta_1 & \cos\theta_1 & 0 \\ 0 & 0 & 0 & 1 \end{bmatrix}$$

$$e^{\hat{\xi}_2\theta_2} = \begin{bmatrix} e^{\hat{\omega}_2\theta_2} & (\mathbf{I} - e^{\hat{\omega}_2\theta_2})(\omega_2 \times \mathbf{v}_2) \\ \mathbf{0} & 1 \end{bmatrix}$$

$$= \begin{bmatrix} 1 & 0 & 0 & 0 \\ 0 & \cos\theta_2 & -\sin\theta_2 & l\sin\theta_2 \\ 0 & \sin\theta_2 & \cos\theta_2 & l(1-\cos\theta_2) \\ 0 & 0 & 0 & 1 \end{bmatrix}$$

$$e^{\hat{\xi}_3\theta_3} = \begin{bmatrix} e^{\hat{\omega}_3\theta_3} & (\mathbf{I} - e^{\hat{\omega}_3\theta_3})(\omega_3 \times \mathbf{v}_3) \\ \mathbf{0} & 1 \end{bmatrix}$$

$$= \begin{bmatrix} 1 & 0 & 0 & 0 \\ 0 & \cos\theta_3 & -\sin\theta_3 & 2l\sin\theta_3 \\ 0 & \sin\theta_3 & \cos\theta_3 & 2l(1-\cos\theta_3) \\ 0 & 0 & 0 & 1 \end{bmatrix} \tag{A.20}$$

The forward kinematics of the tool frame T can be obtained as

$$\mathbf{g}_{ST}(\theta) = e^{\hat{\xi}_1\theta_1} e^{\hat{\xi}_2\theta_2} e^{\hat{\xi}_3\theta_3} \mathbf{g}_{ST}(0)$$

$$= \begin{bmatrix} 1 & 0 & 0 & 0 \\ 0 & -\sin\theta_{123} & -\cos\theta_{123} & -l(\sin\theta_{123} + \sin\theta_{12} + \sin\theta_1) \\ 0 & \cos\theta_{123} & -\sin\theta_{123} & l(\cos\theta_{123} + \cos\theta_{12} + \cos\theta_1) \\ 0 & 0 & 0 & 1 \end{bmatrix} \tag{A.21}$$

where $\theta_{123} = \theta_1 + \theta_2 + \theta_3$ and $\theta_{12} = \theta_1 + \theta_2$.

We now obtain the twist of the tool frame T. It follows from Eq. (A.17) that

$$\xi_1 = \xi_1 = \begin{bmatrix} 0 \\ 0 \\ 0 \\ 1 \\ 0 \\ 0 \end{bmatrix}$$

$$\xi_2' = \mathrm{Ad}_{e^{\hat{\xi}_1\theta_1}}\, \xi_2 = \begin{bmatrix} 0 \\ l\cos\theta_1 \\ l\sin\theta_1 \\ 1 \\ 0 \\ 0 \end{bmatrix}$$

$$\xi_3' = \mathrm{Ad}_{e^{\hat{\xi}_1\theta_1}e^{\hat{\xi}_2\theta_2}}\xi_3 = \begin{bmatrix} 0 \\ l(\cos\theta_{12} + \cos\theta_1) \\ l(\sin\theta_{12} + \sin\theta_1) \\ 1 \\ 0 \\ 0 \end{bmatrix} \tag{A.22}$$

Hence, the twist of the tool frame T is given by

$$\xi_{ST} = \mathbf{J}_{ST}\dot{\theta} \tag{A.23}$$

where

$$\mathbf{J}_{ST} = [\xi_1\ \xi_2'\ \xi_3'] = \begin{bmatrix} 0 & 0 & 0 \\ 0 & l\cos\theta_1 & l(\cos\theta_{12} + \cos\theta_1) \\ 0 & l\sin\theta_1 & l(\sin\theta_{12} + \sin\theta_1) \\ 1 & 1 & 1 \\ 0 & 0 & 0 \\ 0 & 0 & 0 \end{bmatrix}, \quad \dot{\theta} = \begin{bmatrix} \dot{\theta}_1 \\ \dot{\theta}_2 \\ \dot{\theta}_3 \end{bmatrix}$$

This completes the forward position and velocity kinematics of the $3R$ manipulator based on the POE formula.

Inverse Position and Velocity Kinematics

Inverse position kinematics is to obtain the joint angles given the position and orientation $\mathbf{g}_{ST}$ of the tool frame T w.r.t. the frame S, and inverse velocity kinematics is to obtain the joint angle rates given the twist ξ_{ST} of the tool frame T w.r.t. the frame S.

It can be seen from Eq. (A.21) that the fingertip O' can only move in the plane spanned by the $\mathbf{j}$ and $\mathbf{k}$-axes, as shown in Fig. A.3.

Suppose the position vector of the point O' is given as $(0, y, z)$ and the $\mathbf{k}'$-axis makes an angle θ with the $\mathbf{k}$-axis. The joint angles θ_1–θ_3 are referenced from their reference lines respectively, as shown

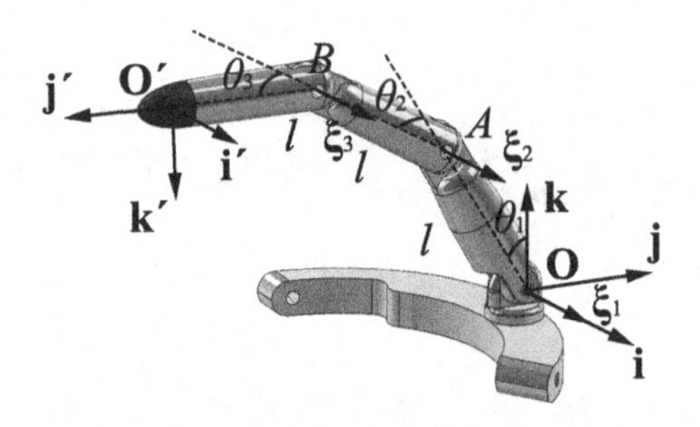

Fig. A.3: The joint angles of the $3R$ manipulator

in Fig. A.3. It can be seen straightforwardly that

$$\theta_{123} = \theta - \frac{\pi}{2} \tag{A.24}$$

The following equations exist for the position vector of the point O' from Eq. (A.21):

$$\begin{aligned} -l(\sin\theta_{123} + \sin\theta_{12} + \sin\theta_1) &= y \\ l(\cos\theta_{123} + \cos\theta_{12} + \cos\theta_1) &= z \end{aligned} \tag{A.25}$$

It is straightforward to solve Eqs. (A.24) and (A.25) for θ_1–θ_3. We skip the steps here.

Suppose the twist ξ_{ST} of the frame T w.r.t. the frame S is given as

$$\xi_{ST} = \begin{bmatrix} 0 \\ v_y \\ v_z \\ \dot{\theta} \\ 0 \\ 0 \end{bmatrix} \tag{A.26}$$

Equating Eq. (A.23)–Eq. (A.26) gives the following three equations:

$$\begin{aligned} l\cos\theta_1\dot{\theta}_2 + l(\cos\theta_{12} + \cos\theta_1)\dot{\theta}_3 &= v_y \\ l\sin\theta_1\dot{\theta}_2 + l(\sin\theta_{12} + \sin\theta_1)\dot{\theta}_3 &= v_z \\ \dot{\theta}_1 + \dot{\theta}_2 + \dot{\theta}_3 &= \dot{\theta} \end{aligned} \tag{A.27}$$

It is straightforward to solve Eq. (A.27) for the joint angle rates $\dot{\theta}_1$–$\dot{\theta}_3$. We skip the steps here.

This completes solving the inverse position and velocity kinematics of the $3R$ manipulator. □

Bibliography

Agrachev, A. A. and Sachkov, Y. L. (1999). An intrinsic approach to the control of rolling bodies, in *IEEE Conference on Decision and Control*, Vol. 1, ISBN 0191-2216, pp. 431–435 vol.1, doi:10.1109/CDC.1999.832815.

Angeles, J. (1988). *Rational Kinematics* (Springer, New York), ISBN 978-1-4612-8400-0, doi:https://doi.org/10.1007/978-1-4612-3916-1.

Ardema, M. D. (2004). *Analytical Dynamics: Theory and Applications* (Springer).

Aydinalp, M., Kazaz, M., and Uğurlu, H. H. (2018). The forward kinematics of rolling contact of spacelike surfaces in Lorentzian 3-space, *Cumhuriyet Science Journal* **39**, 4, pp. 882–899.

Bayar, A., Altıntaş, A., Ekmekçi, S., and Akça4, Z. (2018). Some properties of parabolas whose vertices are on sides of orthic triangle and foci are orthocenter, in *16th International Geometry Symposium*, Manisa, Turkey.

Blažič, S. (2011). A novel trajectory-tracking control law for wheeled mobile robots, *Robotics and Autonomous Systems* **59**, 11, pp. 1001–1007.

Bullo, F. and Lewis, A. D. (2005). *Geometric Control of Mechanical Systems: Modeling, Analysis, and Design for Simple Mechanical Control Systems* (Springer).

Cai, C. and Roth, B. (1986). On the planar motion of rigid bodies with point contact, *Mechanism and Machine Theory* **21**, 6, pp. 453–466.

Carmo, M. P. (1976). *Differential Geometry of Curves and Surfaces* (Prentice-Hall, Englewood Cliffs).

Cartan, E. (2002). *Riemannian Geometry in an Orthogonal Frame* (World Scientific Press, Singapore).

Cartan, H. (1996). *Differential Forms* (Dover Publisher, New York).

Casey, J. and Lam, V. C. (1986). A tensor method for the kinematical analysis of systems of rigid bodies, *Mechanism and Machine Theory* **21**, 1, pp. 87–97, doi: https://doi.org/10.1016/0094-114X(86)90032-7; http://www.sciencedirect.com/science/article/pii/0094114X86900327.

Chelouah, A. and Chitour, Y. (2003). On the motion planning of rolling surfaces, *Forum Mathematicum* **15**, 5, p. 727, doi:10.1515/form.2003.039; https://www.degruyter.com/view/j/form.2003.15.issue-5/form.2003.039/form.2003.039.xml.

Chen, C.-H. (1997). Conjugation form of motion representation and its conversion formulas, *Mechanism and Machine Theory* **32**, 6, pp. 765–774, doi:https://doi.org/10.1016/S0094-114X(97)83008-X; http://www.sciencedirect.com/science/article/pii/S0094114X9783008X.

Chern, S., Chen, W., and Lam, K. (1999). *Lectures on Differential Geometry* (World Scientific), ISBN 9789810241827; https://books.google.com.au/books?id=Wm8dQ9sHl4oC.

Chitour, Y., Marigo, A., and Piccoli, B. (2005). Quantization of the rolling-body problem with applications to motion planning, *Systems and Control Letters* **54**, 10, pp. 999–1013.

Cole, A., Hauser, J. E., and Sastry, S. (1989). Kinematics and control of multifingered hands with rolling contact, *IEEE Transactions on Automation Control* **34**, 4, pp. 398–404.

Cui, L. (2010). *Differential Geometry Based Kinematics of Sliding-Rolling Contact and Its Use for Multifingered Hands*, Thesis. King's college London, London, UK.

Cui, L., Cupcic, U., and Dai, J. S. (2014). An optimization approach to teleoperation of the thumb of a humanoid robot hand: Kinematic mapping and calibration, *Journal of Mechanical Design* **136**, 9, pp. 091005–091005–7.

Cui, L. and Dai, J. S. (2010). A Darboux-frame-based formulation of spin–rolling motion of rigid objects with point contact, *IEEE Transactions on Robotics* **26**, 2, pp. 383–388, doi:10.1109/TRO.2010.2040201.

Cui, L. and Dai, J. S. (2012). Reciprocity-based singular value decomposition for inverse kinematic analysis of the metamorphic multifingered hand, *ASME Journal of Mechanisms and Robotics* **4**, 3, pp. 034502–034502–6.

Cui, L. and Dai, J. S. (2015). A polynomial formulation of inverse kinematics of rolling contact, *Journal of Mechanisms and Robotics* **7**, 4, pp. 041003–041003–9, doi:10.1115/1.4029498.

Cui, L. and Dai, J. S. (2018). *Rolling Contact in Kinematics of Multifingered Robotic Hands* (Springer International Publishing, Cham), ISBN 978-3-319-56802-7, pp. 217–224, doi:10.1007/978-3-319-56802-7_23; https://doi.org/10.1007/978-3-319-56802-7_23.

Cui, L., Sun, J., and Dai, J. S. (2017). In-hand forward and inverse kinematics with rolling contact, *Robotica*, **35**, 12 pp. 1–19.

Cui, L., Wang, D., and Dai, J. S. (2009). Kinematic geometry of circular surfaces with a fixed radius based on Euclidean invariants, *ASME Journal of Mechanical Design* **131**, 10, pp. 101009–8, doi:10.1115/1.3212679; http://link.aip.org/link/?JMD/131/101009/1.

Dafle, N. C., Rodriguez, A., Paolini, R., Tang, B., Srinivasa, S. S., Erdmann, M., Mason, M. T., Lundberg, I., Staab, H., and Fuhlbrigge, T. (2014). Extrinsic dexterity: In-hand manipulation with external forces, in *IEEE International Conference on Robotics and Automation*, pp. 1578–1585.

Dai, J. S., Wang, D., and Cui, L. (2009). Orientation and workspace analysis of the multifingered metamorphic hand-metahand, *IEEE Transactions on Robotics* **25**, 4, pp. 942–947, doi:10.1109/TRO.2009.2017138, 10.1109/TRO.2009.2017138.

De Schutter, J. (2009). Invariant description of rigid body motion trajectories, *Journal of Mechanisms and Robotics* **2**, 1, pp. 011004–011004–9, doi:10.1115/1.4000524; http://dx.doi.org/10.1115/1.4000524.

Dixon, W. E., Dawson, D. M., Zergeroglu, E., and Behal, A. (2001). *Nonlinear Control of Wheeled Mobile Robots* (Springer-Verlag New York, Inc.). ISBN 1852334142.

Dixon, W. E., de Queiroz, M. S., Dawson, D. M., and Flynn, T. J. (2004). Adaptive tracking and regulation of a wheeled mobile robot with controller/update law modularity, *IEEE Transactions on Control Systems Technology* **12**, 1, pp. 138–147.

Exner, C. (1989). Development of hand functions, *Occupational Therapy for Children*, **1**, 1, pp. 235–259.

Furukawa, N., Namiki, A., Taku, S., and Ishikawa, M. (2006). Dynamic regrasping using a high-speed multifingered hand and a high-speed vision system, in *IEEE International Conference on Robotics and Automation* ISBN 0780395050, pp. 181–187.

Gauss, K. F. and Pesic, P. (2005). *General Investigations of Curved Surfaces* (Courier Corporation).

Harada, K., Kawashima, T., and Kaneko, M. (2002). Rolling based manipulation under neighborhood equilibrium, *The International Journal of Robotics Research* **21**, 5–6, pp. 463–474, doi:10.1177/027836402761393487; https://journals.sagepub.com/doi/abs/10.1177/027836402761393487.

Javadi A. A. H. and Mojabi, P. (2004). Introducing glory: A novel strategy for an omnidirectional spherical rolling robot, *Journal of Dynamic Systems, Measurement, and Control* **126**, 3, pp. 678–683, http://link.aip.org/link/?JDS/126/678/1.

Joshi, V. A., Banavar, R. N., and Hippalgaonkar, R. (2010). Design and analysis of a spherical mobile robot, *Mechanism and Machine Theory* **45**, 2, pp. 130–136, doi:10.1016/j.mechmachtheory.2009.04.003; http://dx.doi.org/10.1016/j.mechmachtheory.2009.04.003.

Kerr, J. and Roth, B. (1986). Analysis of multifingered hands, *The International Journal of Robotics Research* **4**, 4, pp. 3–17.

Kim, Y.-M., Ahn, S.-S., and Lee, Y.-J. (2010). Kisbot: New spherical robot with arms, in *The 10th WSEAS International Conference on Robotics, Control and Manufacturing Technology* (World Scientific and Engineering Academy and Society (WSEAS), 1844611), pp. 63–67.

Kirson, Y. and Yang, A. T. (1978). Instantaneous invariants in three-dimensional kinematics, *Journal of Applied Mechanics* **45**, 2, pp. 409–414, doi:10.1115/1.3424310; http://dx.doi.org/10.1115/1.3424310.

Kiss, B., Lévine, J., and Lantos, B. (2002). On motion planning for robotic manipulation with permanent rolling contacts, *The International Journal of Robotic Research* **21**, 5–6, pp. 443–461.

Legnani, G., Casolo, F., Righettini, P., and Zappa, B. (1996). A homogeneous matrix approach to 3D kinematics and dynamics — I. theory, *Mechanism and Machine Theory* **31**, 5, pp. 573–587, doi:https://doi.org/10.1016/0094-114X(95)00100-D; http://www.sciencedirect.com/science/article/pii/0094114X9500100D.

Li, T. and Liu, W. (2011). Design and analysis of a wind-driven spherical robot with multiple shapes for environment exploration, *Journal of Aerospace Engineering* **24**, 1, pp. 135–135, doi:10.1061/(ASCE)AS.1943-5525.0000063; http://ascelibrary.org/aso/resource/1/jaeeez/v24/i1/p135_s1?view=fulltext.

Li, Z. X. and Canny, J. (1990). Motion of two rigid bodies with rolling constraint, *IEEE Transactions on Robotics and Automation* **6**, 1, pp. 62–72.

Lipschutz, M. M. (1969). *Schaum's Outline of Theory and Problems of Differential Geometry*, Schaum's Outline Series (McGraw-Hill Book Company, New York).

Ma, R. R. and Dollar, A. M. (2013). Linkage-based analysis and optimization of an underactuated planar manipulator for in-hand manipulation, *Journal of Mechanisms and Robotics* **6**, 1, pp. 011002–011002, doi:10.1115/1.4025620; http://dx.doi.org/10.1115/1.4025620.

Maekawa, H., Tanie, K., and Komoriya, K. (1997). Kinematics, statics and stiffness effect of 3D grasp by multifingered hand with rolling contact at the fingertip, in *International Conference on Robotics and Automation*, Vol. 1, pp. 78–85 doi:10.1109/ROBOT.1997.620019.

Marigo, A. and Bicchi, A. (2000). Rolling bodies with regular surface: Controllability theory and application, *IEEE Transactions on Automatic Control* **45**, 9, pp. 1586–1599.

Montana, D. J. (1988). The kinematics of contact and grasp, *The International Journal of Robotics Research* **7**, 3, pp. 17–32.

Montana, D. J. (1992). Contact stability for two-fingered grasps, *IEEE Transactions on Robotics and Automation* **8**, 4, pp. 421–430.

Montana, D. J. (1995). The kinematics of multi-fingered manipulation, *IEEE Transactions on Robotics and Automation* **11**, 4, pp. 491–503.

Murray, Richard M. and Li, Zexiang and Sastry, S. Shankar (1994). *A Mathematical Introduction to Robotic Manipulation*, CRC Press, Boca Raton, USA.

Neimark, I. I. and Fufaev, N. A. (1972). *Dynamics of Nonholonomic Systems* (American Mathematical Society).

Nomizu, K. (1978). Kinematics and differential geometry of submanifolds: Rolling a ball with a prescribed locus of contact, *Tohoku Mathematical Journal* **30**, 4, pp. 623–637.

Pehoski, C., Henderson, A., and Tickle-Degnen, L. (1997). In-hand manipulation in young children: Rotation of an object in the fingers, *American Journal of Occupational Therapy* **51**, 7, pp. 544–552.

Remond, C., Perdereau, W., and Drouin, M. (2002). A hierarchical multi-fingered hand control structure with rolling contact compensation, in *IEEE International Conference on Robotics and Automation*, Vol. 4, pp. 3731–3736, doi:10.1109/ROBOT.2002.1014292.

Rosenberg, M. R. (1977). *Analytical Dynamics of Discrete Systems* (Plenum Press).

Roth, B. (2005). Finding geometric invariants from time-based invariants for spherical and spatial motions, *ASME Transactions: Journal of Mechanical Design* **127**, 2, pp. 227–231, http://link.aip.org/link/?JMD/127/227/1.

Roth, B. (2015). On the advantages of instantaneous invariants and geometric kinematics, *Mechanism and Machine Theory* **89**, pp. 5–13, doi:https://doi.org/10.1016/j.mechmachtheory.2014.10.009; http://www.sciencedirect.com/science/article/pii/S0094114X14002560.

Sarkar, N., Kumar, V., and Yun, X. (1996). Velocity and acceleration equations for three-dimensional contact, *Journal of Applied Mechanics* **63**, 4, pp. 974–984.

Sarkar, N., Yun, X., and Kumar, V. (1997). Control of contact interactions with acatastatic nonholonomic constraints, *The International Journal of Robotic Research* **16**, 3, pp. 357–374.

Sasaki, S. (1955). *Differential Geometry (in Japanese)* (Kyorlitsu Press).

Selig, J. M. (2005). *Geometric fundamentals of robotics*, (Springer), pp. 398–398, ISBN=0387208747 http://books.google.com/books?hl=en&lr=&id=9FljXoISr8AC&pgis=1.

Soetanto, D., Lapierre, L., and Pascoal, A. (2003). Adaptive, non-singular path-following control of dynamic wheeled robots, in *IEEE Conference on Decision and Control*, Vol. 2 (IEEE), ISBN 0780379241, pp. 1765–1770.

Sudsang, A. and Phoka, T. (2003). Regrasp planning for a 4-fingered hand manipulating a polygon, in *IEEE International Conference on Robotics and Automation*, Vol. 2 (IEEE), ISBN 0780377362, pp. 2671–2676.

Tchon, K. (2002). Repeatability of inverse kinematics algorithms for mobile manipulators, *IEEE Transactions on Automatic Control* **47**, 8, pp. 1376–1380.

Tchon, K. and Muszynski, R. (1998). Singular inverse kinematic problem for robotic manipulators: A normal form approach, *IEEE Transactions on Robotics and Automation* **14**, 1, pp. 93–104.

Veldkamp, G. R. (1967). Canonical systems and instantaneous invariants in spatial kinematics, *Journal of Mechanisms* **2**, 3, pp. 329–388, doi:https://doi.org/10.1016/0022-2569(67)90006-7; http://www.sciencedirect.com/science/article/pii/0022256967900067.

Wang, D., Liu, J., and Da Zhun, X. (1997a). Kinematic differential geometry of a rigid body in spatial motion–I. a new adjoint approach and instantaneous properties of a point trajectory in spatial kinematics, *Mechanism and Machine Theory* **32**, 4, pp. 419–432, doi:https://doi.org/10.1016/S0094-114X(96)00075-4; http://www.sciencedirect.com/science/article/pii/S0094114X96000754.

Wang, D., Liu, J., and Da Zhun, X. (1997b). Kinematic differential geometry of a rigid body in spatial motion–II. a new adjoint approach and instantaneous properties of a line trajectory in spatial kinematics, *Mechanism and Machine Theory* **32**, 4, pp. 433–444, doi:https://doi.org/10.1016/S0094-114X(96)00076-6; http://www.sciencedirect.com/science/article/pii/S0094114X96000766.

Wang, D. and Wang, W. (2015). *Kinematic Differential Geometry and Saddle Synthesis of Linkages* (John Wiley & Sons Incorporated).

Wang, D. and Xiao, D. (1993). Distribution of coupler curves for crank-rocker linkages, *Mechanism and Machine Theory* **28**, 5, pp. 671–684, doi:https://doi.org/10.1016/0094-114X(93)90007-I; http://www.sciencedirect.com/science/article/pii/0094114X9390007I.

Wang, D. L., Liu, J., and Xiao, D. Z. (1997c). Kinematic differential geometry of a rigid body in spatial motion–III. distribution of characteristic lines in the moving body in spatial motion, *Mechanism and Machine Theory* **32**, 4, pp. 445–457, doi:https://doi.org/10.1016/S0094-114X(96)00077-8; http://www.sciencedirect.com/science/article/pii/S0094114X96000778.

Xu, Y. (1999). Dynamic mobility with single-wheel configuration, *The International Journal of Robotics Research* **18**, 7, pp. 728–738, doi:10.1177/02783649922066538; http://ijr.sagepub.com/cgi/content/abstract/18/7/728.

Yousef, H., Boukallel, M., and Althoefer, K. (2011). Tactile sensing for dexterous in-hand manipulation in robotics—A review, *Sensors and Actuators A: Physical* **167**, 2, pp. 171–187, doi:http://dx.doi.org/10.1016/j.sna.2011.02.038; http://www.sciencedirect.com/science/article/pii/S0924424711001105.

Index

Printed in the United States
By Bookmasters